Guidelines for

Landscape

and Visual Impact

Assessment

OTHER TITLES FROM E & FN SPON

Amenity Landscape Management
A resources handbook
R. Cobham

Beazley's Design and Detail of the Space Between Buildings
A. Pinder and A. Pinder

Countryside Recreation, Access and Land Use Policy
N. Curry

Countryside Management
P. Bromley

Elements in Visual Design in the Landscape
S. Bell

Environmental Planning for Site Development
A. R. Beer

Fungal Diseases of Amenity Turf Grasses
J. Drew Smith, N. Jackson and R. Woodhouse

The Golf Course
F.W. Hawtree

Grounds Maintenance
A contractor's guide to competitive tendering
P. Sayers

Planning and the Heritage
Policy and procedures
M. Ross

Recreational Land Management
W. Seabrooke and C. W. N. Miles

Spon's Landscape and External Works Price Book
Derek Lovejoy and Partners and Davis Langdon & Everest

Spon's Landscape Handbook (3rd edition)
Derek Lovejoy and Partners

Tree Form, Size and Colour
B. J. Gruffydd

Trees in the Urban Landsacpe
A. Bradshaw, B. Hunt and T. Walmsley

For more information on these and other titles please contact:
The Promotion Department, E & FN Spon, 2-6 Boundary Row, London, SE1 8HN
Telephone 0171 865 0066.

Guidelines for

Landscape

and Visual Impact

Assessment

INSTITUTE OF
ENVIRONMENTAL
ASSESSMENT

THE
LANDSCAPE
INSTITUTE

E & FN SPON

An Imprint of Chapman & Hall

London • Glasgow • Weinheim • New York • Tokyo • Melbourne • Madras

Published by E & FN Spon, an imprint of Chapman & Hall,
2–6 Boundary Row, London SE1 8HN, UK

Chapman & Hall, 2–6 Boundary Row, London SE1 8HN, UK

Blackie Academic & Professional, Wester Cleddens Road, Bishopbriggs,
Glasgow G64 2NZ, UK

Chapman & Hall GmbH, Pappelallee 3, 69469 Weinheim, Germany

Chapman & Hall USA, 115 Fifth Avenue, New York, NY10003, USA

Chapman & Hall Japan, ITP-Japan, Kyowa Building, 3F, 2-2-1 Hirakawacho,
Chiyoda-ku, Tokyo 102, Japan

Chapman & Hall Australia, 102 Dodds Street, South Melbourne,
Victoria 3205, Australia

Chapman & Hall India, R. Seshadri, 32 Second Main Road, CIT East,
Madras 600 035, India

First edition 1995

© 1995 Institute of Environmental Assessment and The Landscape Institute

Printed and bound in Hong Kong

ISBN 0 419 20380 X

A Catalogue record for this book is available from the British Library

Contents

Foreword

The landscape is a vital part of our natural and man-made environment and is considered by many to be one of the most important components of a healthy, enjoyable life. During the latter part of this century we have witnessed its vulnerability in the face of economic growth and social change. Often, we have failed adequately to predict, recognise or deal with the impact of new development. The result has been serious erosion of the character and quality of many of our urban and rural landscapes.

In planning for future prosperity, more effort must be put into safeguarding the quality of the environment, not only for our own health and well-being, but also for the sake of future generations. This growing emphasis on 'sustainable development' - the balance between social concerns, economic development and the environment - implies a strong need to integrate issues of landscape conservation and enhancement into the development process.

Therefore, it is appropriate that the Institute of Environmental Assessment and the Landscape Institute should join forces to present good practice guidelines for landscape and visual impact assessment. The guidelines are intended to assist in predicting and judging the significance of the effects that new development may have upon landscape character and visual amenity. More importantly, they require a commitment to prevent and reduce adverse impacts and to look for positive opportunities for environmental enhancement.

In this way, economic well-being and environmental quality can go hand-in-hand. The guidelines represent the two institutes' preferred approach to achieving these dual objectives through sound landscape assessment, planning and design. We very much hope that they can be applied both formally, for those projects where authorisation procedures require an environmental statement, and more generally, for all developments that may affect sensitive areas of rural or urban landscape.

Alex Novell FLI, Chairman Landscape Institute Environment Committee

Karl Fuller, Institute of Environmental Assessment

Preface

These good practice guidelines for landscape and visual impact assessment were initiated jointly by the Institute of Environmental Assessment and the Landscape Institute. They have been prepared by a Working Party comprising representatives from a wide range of interests including those who commission environmental assessment work, those who undertake it, and those who are responsible for reviewing environmental statements and reports. It is intended that this document will be periodically reviewed and updated in the light of evolving practice and legislation.

The Working Party members were:

Alex Novell (Chairman)	Novell Tullett
Gary Coulson	Terence O'Rourke plc
Merrick Denton-Thompson	Hampshire County Council
David Eagar	Countryside Council for Wales
Karl Fuller	Institute of Environmental Assessment
Rebecca Hughes	Scottish Natural Heritage
Brian Lewis	Department of Transport
Julie Martin/Clare Harper	Cobham Resource Consultants (Editors)
Rick Minter	Countryside Commission
Nicholas Pearson	Nicholas Pearson Associates
Conor Skehan	Irish Landscape Institute
Martin Slater	Institute of Environmental Assessment
Carly Tinkler	Singleton Landscape
Richard Tisdall	Bovis Homes Technical Services

The production of the guidelines was funded by the Countryside Commission, the Countryside Council for Wales and Scottish Natural Heritage, whose support is gratefully acknowledged. The Working Party would also like to thank Rob Benington of Singleton Landscape, Steven Warnock (consultant) and Susan Griffiths (consultant) for their help in drafting the guidelines.

The guidelines represent the consensus views of Working Party members, but not necessarily the views of their employers.

Acknowledgements

The Institute of Environmental Assessment gratefully acknowledges the illustrations contributed by the following organisations: Bovis Homes Technical Services (p49); BP Exploration (p14); Cobham Resource Consultants (p39, p42, p44, p57); DK Symes Associates & Southern Water Services Ltd (p33); Environmental Resources Management (p71); Granada Hospitality Ltd (p51, p69); Hamsphire County Council (p9, p15, p35, p78); John Feltwell/Wildlife Matters (p18, p34); Karl Fuller (p10, p15); Kent County Council (p42); Leeds Metropolitan University, Landscape Design and Community Unit (p16); Martin Jones - Photographer (p16); Manchester Airport Plc (p29, p31); Novell Tullett (p20); Rail Link Project (p44); Rank Holidays & Hotels Developments Ltd (p64, p66, p67); RSK Environment Ltd (p71, p72); Rugby Cement (p79); The Highways Agency (p79); Tyseley Waste Disposal Ltd (p41); Winfrith Technology Centre (p59).

Summary

Environmental Assessment (EA) has become a widely used tool for aiding decision making. In order to improve the standard of EA the Institute of Environmental Assessment is producing a series of guidelines for assessing different types of impact. These particular guidelines, produced jointly with the Landscape Institute, are designed to set high standards for the scope and content of landscape and visual impact assessments. The guidance is not, however, intended to be an exhaustive manual of methods and techniques.

With the increased emphasis on sustainable development there is a particular need to integrate issues of landscape conservation and enhancement into the development process. Landscape and visual impact assessment is an important component of landscape planning in which the best environmental fit for a development is sought.

Landscape and visual impacts are independent, but related issues. Landscape impacts relate to changes in the fabric, character and quality of the landscape, whilst visual impacts relate to the appearance of these changes. When undertaking a landscape and visual impact assessment it is important to adhere to a number of important principles: describing clearly the methodologies and techniques used; using clearly defined and agreed terminology; maintaining impartiality; drawing upon the advice and opinions of other people; and acknowledging any deficiencies or limitations inherent in the assessment.

Understanding the nature of the proposed development is vital to a landscape and visual impact assessment. This should include a consideration of alternatives and a clear description of the components of the development which will affect the landscape. All stages of the project life-cycle should be addressed; site preparation, construction, operation and decommissioning.

The collection and analysis of baseline information for the environment, in which the proposed project is to be placed, should cover the process of description, classification and evaluation of the landscape resource. Collection of the data will require an initial desk study and essential field studies. All of the potential landscape and visual impacts of the development should be identified, the magnitude of the impacts predicted and the significance of those impacts assessed. Mitigation measures designed to avoid or minimise impacts or enhance the existing landscape should be considered and their likely effectiveness should be described.

EA is an aid to decision making and a medium by which the environmental effects of a project are conveyed to the public. The communication of the information is therefore an important element of the assessment process. In addition to having clear descriptions of the project and its likely impact, the assessment should use illustrative techniques which convey the information effectively.

Consultation with decision making authorities, statutory consultees and the public should be conducted throughout the process. The assessment should not be considered to have finished when development consent is granted. During the implementation of the project, impacts and the effectiveness of mitigation measures should be monitored in order that any unanticipated effects can be dealt with promptly and effectively.

Part One

Introduction

Background

1.1   Environmental Assessment is an environmental management tool which has been in use on an international basis since 1970. It is a process by which the identification, prediction and evaluation of the key impacts of a development is undertaken and the information gathered is used to improve the design of the project and to inform the decision making process. The process is illustrated by the diagram in Appendix 1. In 1985, the European Community (EC) Directive 85/337/EEC [1] on Environmental Assessment (EA) first established a formal requirement for EA of certain categories of public and private projects. The Directive was implemented in the UK in 1988 through a series of statutory instruments known as the EA Regulations (see Appendix 5).

1.2   In more recent moves towards sustainable development, government policy has developed the theme of protection and enhancement of the environment as an integral part of planning for new development [3, 4]. EA is a vital tool for analysing the environmental issues associated with development, for improving the siting, layout and design of a particular scheme and for aiding decision making.

1.3   The growing role of EA within the development process is borne out by research by the Institute of Environmental Assessment (IEA). This shows that in the six years to 1993 at least 1600 formal environmental statements (ESs) were submitted as part of planning and other project authorisation procedures, with many being supplied on a voluntary basis. It is clear that EA is now being used very widely as an aid to the decision making process for many forms of development, not just for those covered by the EC Directive. Its strength is in enabling the best 'environmental fit' between a project and its surroundings, and in helping to determine whether the development is acceptable.

1.4   The standard and content of many ESs prepared to date has been subject to criticism, and reviews of ESs carried out by Manchester University [5] and by the IEA [6] have highlighted significant inadequacies. Concern has also been expressed that the regulations have not always been applied consistently, perhaps due to lack of familiarity with or understanding of good practice. For all these reasons, there is a clear need for sound, reliable and widely accepted advice on best practice for EA.

Aims of the guidelines

1.5   To meet this need, the IEA has embarked upon the preparation of a series of good practice guidelines for assessing different types of environmental impact. The guidance in this document is specifically intended to cover all aspects of landscape and visual impact assessment and has been produced jointly with the Landscape Institute (LI) to ensure that, in particular, they fully meet the requirements of the landscape profession.

1.6   They are aimed at a wide audience, including:

*   developers
*   professional project teams
*   those responsible for contributing to and managing EAs
*   planners and others within local government and the various government agencies responsible for EA review
*   politicians, amenity societies and the general public, who are the main readership for ESs
*   academics and students of landscape design and EA.

1.7   The aim of the guidelines is to set high standards for the scope and content of landscape and visual impact assessments, and to establish certain principles which will ensure integrity and consistency. Some guidance is also given on preferred methods and techniques for different aspects of the assessment process in appendices. However, the guidelines are not intended as a prescriptive set of rules nor as an exhaustive manual of methods and techniques.

1.8   It is recognised that, unlike some other aspects of EA, landscape and visual impact assessment relies less upon measurement than upon experience and judgement; although all do have a part to play. In this context, a structured and consistent approach to the treatment of landscape and visual issues is important. In offering guidance on the establishment of significance in landscape and visual impact assessment, it is necessary to take care to differentiate between judgement on the significance of change, which involves a greater degree of subjective opinion, and the measurement of magnitude of change, which is normally a more objective and quantifiable task. Judgement, should always be supported by clear evidence, reasoned argument and informed opinion. It is recommended that in most instances landscape and visual impact assessments should be carried out and reviewed by suitably qualified landscape professionals, although this is not to say that other professionals may not also have a valid role.

Scope of the guidelines

1.9　　After the Introduction (**Part One**), the guidelines have been organised into three main sections:

1.10　　**Part Two** presents the common principles and philosophy which should underlie the assessment of landscape and visual impacts. It sets out the current policy towards the conservation of our landscape and visual resources and examines the distinction between landscape and visual impacts, the ways in which the landscape's character and appearance may be altered by development, and the various sensitive 'receptors' of landscape and visual impact. Finally, it emphasises the role of landscape and visual impact assessment within the process of sound landscape planning, and establishes a number of common principles that should be applied to every assessment.

1.11　　**Part Three** briefly examines the EA process as a whole and the role of landscape and visual impact assessment within it. It then looks in some detail at the main steps in the process, that is: a description of the development; baseline studies; impact assessment; impact mitigation; and presentation of findings. Examples of good practice are included as illustration of the different stages in the assessment process.

1.12　　**Part Four** deals with consultation, review and implementation of the proposed project and the assessment. The role of consultation with both statutory agencies and the general public is explored, and it is recognised that this can greatly assist assessment and decision making in respect of the development. Advice is also given to the determining authorities who are the recipients of environmental reports and formal ESs. They have an important role in guiding the developer through the landscape and visual impact assessment process; reviewing the assessment's scope, technical content and presentation; and implementing and monitoring landscape conditions. Without this important follow-up action, effective impact mitigation is unlikely to be achieved.

1.13　　The readers attention is drawn to the Glossary of Terms used in the guidelines to be found at the end of Part Four, and the helpful information contained in a number of technical appendices.

Part Two

Good practice in landscape and visual impact assessment

Landscape conservation policy

2.1   The term **landscape** commonly refers to the appearance of the land, including its shape, texture, and colours [7]. It also reflects the way in which these various components combine to create specific patterns and pictures that are distinctive to particular localities. The landscape is not a purely visual phenomenon; it relies heavily on other influences for its character. These include the underlying geology and soils, the topography, archaeology, landscape history, land use, land management, ecology, architecture and cultural associations, all of which can influence the ways in which landscape is experienced and valued.

A townscape in Hampshire, illustrating the urban nature of the landscape. The interrelationship between buildings and open spaces, plants and other elements combine to create the urban landscape

2.2   The landscape is not simply a rural phenomenon. It encompasses the whole of our external environment, whether within villages, towns, cities or in the countryside. The patterns and textures of buildings, streets, open spaces and trees, and their interrelationships within the built environment are an equally important part of our wider landscape heritage.

2.3   Whether urban or rural, landscape is important because it is:

- an essential part of our natural resource base;
- a reservoir of archaeological and historical evidence;

- an environment for plants and animals;
- a resource which evokes sensual, cultural and spiritual responses;
- an important part of our quality of life.

A rural landscape showing different aspects of the landscape resource: landuse and management, ecological features, buildings and landform.

2.4   The value of landscape and visual resources is stressed in recent guidance on environmental and conservation issues in development plans [9, 10] and hence will be reflected in future decisions on major development proposals.

2.5   The emphasis of the recent advice, which has come from both the DoE and the conservation agencies, is that greater heed must be paid in future to the concept of 'sustainable development'.

> The most widely used definition of sustainable development comes from the 1987 Brundtland report, namely: "development that meets the needs of the present generation without compromising the ability of future generations to meet their own needs." [15]

Some of the main principles of sustainable development that have been put forward in the literature are that:

- policies should aspire to a balance among societal, economic and environmental objectives, with an emphasis on managing for future needs rather than just providing for society's immediate and ever increasing demand for natural resources;

- resources should be safeguarded for all, with more attention being given to global and inter-generational effects;

- environmental capacities and thresholds should be identified and recognised, which means that limits need to be set on certain types of development in certain areas;

- key environmental resources should be safeguarded, and other environmental resources should be maintained or enhanced;

- the 'precautionary principle' should be applied where uncertainty surrounds the environmental impacts of particular actions.

2.7   The implications for the assessment of landscape and visual impacts are considerable. There will be growing recognition that development may bring impacts not just for the site itself and its immediate environs, but also for other areas. For instance, there are often important choices associated with the use of materials. The construction of a new road may result in significant associated landscape and visual impacts from quarrying; whereas the re-use of construction waste may actually help solve a landscape problem elsewhere.

2.8   The impacts of our actions on future generations will need to be borne in mind. Developments that result in irreversible damage to important environmental resources, including landscape of recognised international, national, regional or local importance, may need to be rejected in favour of others which bring no such damage. Current landscape and visual resources should be at least maintained, and any losses made good through environmental enhancement which genuinely compensates for any features lost or damaged. There will need to be a new focus on development that enhances rather than depletes our stock of landscape and visual resources.

2.9   The EC Directive and existing guidance on EA from the government [2] and from the Countryside Commission [8] emphasise that landscape issues are relevant to the EA process. This suggests a focus on the appearance of the land and the way in which people respond to landscapes, but at the same time indicates that other factors, including flora and fauna, soil, air, water, climate, material assets and cultural heritage, should be borne in mind as contributors to the landscape.

2.10   Landscape and visual issues often play a prominent role in EA. Unlike less obvious impacts such as changes in groundwater quality, changes in the landscape have a direct, immediate, visible effect upon people's surroundings, and therefore may arouse strong feelings. They may also be used by the public as a focus for a variety of other concerns about the impact of a development. Therefore it is essential that assessment of the landscape and visual impacts of a proposed development is carried out in as measured and controlled a way as possible. This will involve not only careful prediction of the nature and scale of potential changes, but also systematic assessment of the significance of those changes for specific landscape and visual resources.

Understanding landscape and visual impacts

2.11   It must first be understood that landscape impacts and visual impacts are separate, but related. **Figure 2.1** summarises the scope of the two terms and their interrelationship.

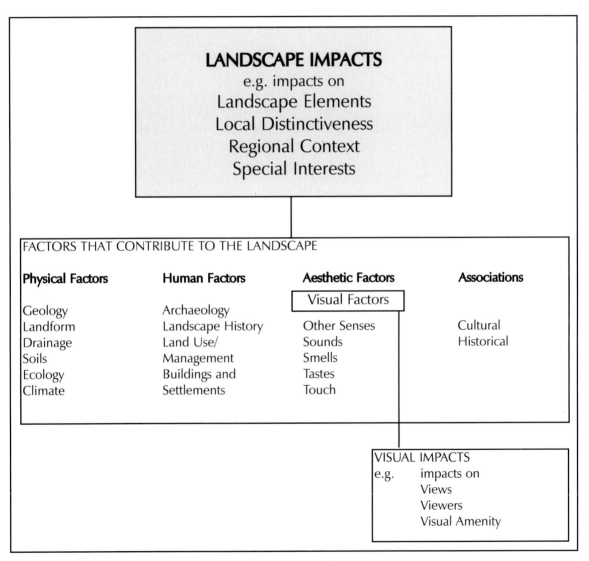

Figure 2.1: The relationship between landscape and visual impacts

2.12   **Landscape impacts** are changes in the fabric, character and quality of the landscape as a result of development. Hence landscape impact assessment is concerned with:

- direct impacts upon specific **landscape elements**;
- more subtle effects upon the overall pattern of elements that gives rise to **landscape character** and **regional and local distinctiveness**;
- impacts upon acknowledged **special interests** or values such as **designated landscapes, conservation sites** and **cultural associations.**

2.13   **Visual impacts** are a subset of landscape impacts. They relate solely to changes in available views of the landscape, and the effects of those changes on people. Hence visual impact assessment is concerned with:

- the **direct impacts** of the development upon views of the landscape through intrusion or obstruction;
- the **reactions of viewers** who may be affected;
- the overall impact on **visual amenity**, which can range from degradation through to enhancement.

2.14   Landscape and visual impacts do not necessarily coincide. Landscape impacts can occur in the absence of visual impacts, for instance where a development is wholly screened from available-views, but nonetheless results in a loss of landscape elements, and landscape character within the site boundary. Similarly, some developments, such as a new communications mast in an industrial area, may have significant visual impacts, but insignificant landscape impacts. However, such cases are very much the exception, and for most developments both landscape and visual impacts will need to be assessed.

Furzey Island in Dorset, part of the BP Wytch Farm Complex. Extensive screening results in a development with minimal visual impact but nevertheless has landscape impact on the landscape character of the area

2.15   Landscape and visual impacts can arise from a variety of **sources.** They can be caused by changes in land use, for example mineral extraction, afforestation and land drainage; by the development of buildings and structures such as power stations, industrial

The visual impact of this development is large in terms of the extent of visibility, but because of its siting in an industrial area, the landscape impacts are insignificant

estates, roads and housing developments; by changes in land management, such as intensification of agricultural use, which can be a vehicle for biological and landscape change; and, less commonly, by changes in production processes and emissions, for instance from chemical, food and textile industry plants.

Landscape and visual impacts occur during the construction stage of a project. The impact from construction housing compounds, plant and machinery and earth movements can be clearly seen in this photograph

2.16   There may be different sources of impact at various stages in the life of a project, that is during construction, operation, decommissioning and restoration.   For example, during construction, sources of impact may include site access, haulage, materials storage and large plant; whereas during operation, permanent infrastructure, structures, lighting and workforce traffic may be the main concerns. In the early stages of an assessment, it is important that all potential sources of impact are systematically identified.

Impact of road lighting and lights from passing cars on a dual carriage-way road

2.17   Impact occurs when landscape or visual resources are affected.   Receptors of landscape and visual impact may include physical and natural landscape and biological resources, special interests and groups of viewers (see Table 2.1).

Children playing in a playground are potentially sensitive to urban landscape change

2.18   The **significance** of a landscape and visual impact *is a function of the sensitivity of the affected landscape and visual receptors and the magnitude of change that they will experience.* The term **sensitivity** incorporates the relative value of the landscape and how tolerant it is to change. By combining magnitude and sensitivity in a systematic fashion,   consistent conclusions on impact significance can be drawn.

<table>
<tr><td colspan="2" align="center">EXAMPLES OF LANDSCAPE AND VISUAL RECEPTORS</td></tr>
<tr><td>Areas of distinctive landscape character</td><td>• Characteristic patterns and combinations of landform and landcover and a sense of place (genius loci)</td></tr>
<tr><td>Valued landscapes</td><td>for example:

• Nationally designated landscapes - National Parks, Heritage Coasts, Areas of Outstanding National Beauty, National Scenic Areas (Scotland)

• Locally designated landscapes - Areas of Great Landscape Value, Conservation Areas

• Areas of local importance - local beauty spots, popular walks and features on Parish Maps</td></tr>
<tr><td>Other conservation interest</td><td>for example:

• Archaeological sites, historic landscapes, important habitats</td></tr>
<tr><td>Specific landscape elements</td><td>for example:

• Coastline, open hilltops, river corridors, woodlands, hedges, walls</td></tr>
<tr><td>Viewers of the landscape</td><td>for example:

• Residents, tourists, visitors, ramblers</td></tr>
</table>

Table 2.1
Landscape and visual receptors

18

2.19   Impacts may be beneficial or adverse, direct or indirect, temporary or permanent, single or cumulative and of course may vary in their magnitude and significance.   These aspects are examined in more detail in Part 3.

This photograph illustrates the effect of an accumulation of successive housing developments in North Cardiff during the 1970s to the 1990s

Landscape planning

2.20   Landscape and visual impact assessment is an important component of landscape planning and is part of a process which seeks to provide the best 'environmental fit' for a development. This is an iterative process which requires the co-operation of the entire project planning team.

2.21   The landscape planing process involves the following steps:

• understand the nature of the receiving environment;

• identify and map all relevant environmental data, including opportunities and constraints;

• adjust the location, layout and design of the development for the best practical environmental fit for each option;

• prepare land use/landscape strategies to best avoid constraints, mitigate impact and take advantage of environmental opportunities for each of the options available;

• compare the options, choosing those with least adverse environmental effect;

• develop a landscape masterplan for the preferred option or options to illustrate how the landscape strategy will work in practice;

• prepare a Landscape and Visual Impact Assessment report.

2.22   Landscape and visual impact assessments are not usually exercises that are carried out in isolation. They are often only one part of a wider assessment of environmental impact that may arise from a proposed development, and it will be important to ensure that the scope and conclusions, particularly on impact significance, are consistent with the overall appraisal framework. Nevertheless the report resulting from a landscape and visual impact assessment should be a 'stand-alone' document. The information may then be integrated into a formal Environmental Statement, or alternatively, it may be used to accompany a planning application in cases where a formal EA has not been undertaken.

The components of the proposed store development have been arranged to achieve the best available fit into its environment

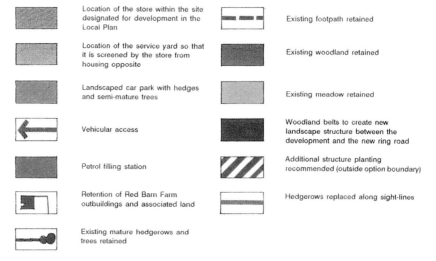

Location of the store within the site designated for development in the Local Plan

Location of the service yard so that it is screened by the store from housing opposite

Landscaped car park with hedges and semi-mature trees

Vehicular access

Petrol filling station

Retention of Red Barn Farm outbuildings and associated land

Existing mature hedgerows and trees retained

Existing footpath retained

Existing woodland retained

Existing meadow retained

Woodland belts to create new landscape structure between the development and the new ring road

Additional structure planting recommended (outside option boundary)

Hedgerows replaced along sight-lines

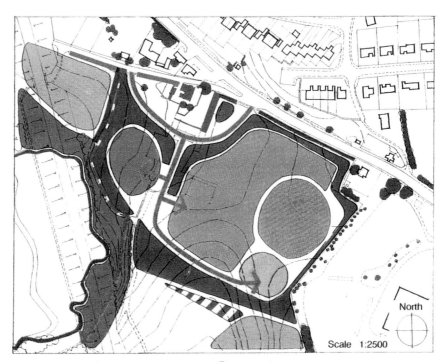

North

Scale 1:2500

Landscape Structure

2.23   Equally, it must be remembered that the assessment will have been instigated by a client who, whilst needing to fulfil the requirements of the DoE or local authority, nevertheless has to undertake the development within a range of constraints, which may be:

- geographical;
- practical;
- financial;
- within a certain time scale.

2.24   The assessor should be aware of all constraints that apply within the context of the overall development. This will include a general understanding of the business or industry to which the assessment is related. Proposals for mitigation should be set within a realistic framework, particularly with regard to cost, and there should be an understanding of the cost/benefit implications of the various proposals. If there are a range of equally effective measures, the most cost effective should be proposed.

2.25   The assessor should use consistently high professional standards both in carrying out the assessment and in making recommendations. Proposals which are most likely to find their way successfully into planning conditions, or an Agreement to be carried out by the developer and be successful in the long term, are those which meet the environmental objectives within development constraints. In essence, mitigation measures should be sound, practicable and cost-effective.

General principles to follow

2.26   The objective of presenting the basic principles of good practice in landscape and visual impact assessment is to ensure:

• **consistency** between landscape and visual impact assessments prepared by different organisations and individuals;

• **thorough and complete coverage** of all potentially significant impacts;

• **reliability** in establishing environmental sensitivities and in predicting impact magnitude;

• **credibility** in the assessment of impact significance, through an appropriate balance of fact and judgement, the clear presentation of all methods, data and logic involved in the assessment process;

• **effectiveness** in avoiding, reducing or ameliorating significant adverse impacts, and in communicating the results of the assessment to decision-makers and the public.

2.27   A number of general principles for good practice have arisen from recent monitoring and review of EA in the UK [5, 6]. Some of the more important lessons that have emerged should be applied to landscape and visual impact assessment:

• **describe clearly the methodology** and the specific techniques that have been used, so that the procedure is replicable and the results can be understood by a lay person;

• **use clearly defined and agreed terminology,** particularly when defining the sensitivity of landscape and visual resources, and the magnitude and significance of predicted impacts;

• **be as impartial as possible,** by distinguishing clearly between objective fact and subjective judgement, and by stating the basis upon which judgements are made;

- **draw upon the advice and opinions of others,** for example, in relation to special interests or values such as archaeology, ecology and the built environment;

- **organise and structure the assessment** to focus upon the key issues of relevance to decision making;

- **openly acknowledge any deficiencies** or limitations of data, techniques or resources that may have constrained the assessment;

- **apply the 'worst case situation' and the 'precautionary principle'** where appropriate, for instance in relation to seasonal or unknown effects.

Part Three

Guidelines for landscape and visual impact assessment

The assessment process

3.1   The main steps in the process of landscape and visual assessment closely mirror the sequence of events which characterise the EA process as a whole (see Appendix 1). The following brief discussion provides an overview of the process and outlines the main points of emphasis as background to the more detailed examination of each stage in the process which then follows.

3.2   The decision on the **need for an** EA will follow guidance provided by the EA Regulations and documents such as the DoE's publication Environmental Assessment: A Guide to the Procedures [2]. Often, a formal EA may not be required, and a series of environmental reports covering key issues will form the basis of a voluntary assessment.

3.3   The **scope and content** of the landscape and visual impact assessment should be established at an early stage. A review of the planning and policy context affecting landscape and visual resources, together with informal discussions and consultations with the determining authority, statutory consultees, amenity bodies, conservation groups and local people, will all help to identify key issues and highlight local sources of information and expertise.

3.4   In setting the scope of the study, the terms of reference, methodology and assessment techniques to be applied, should be clarified. This should involve discussion within the study team and also with the determining authority and the relevant statutory consultees. Agreement may usefully cover:

*   characteristics of the proposed development;
*   limits of the study area;
*   key issues to be addressed;
*   level of detail required for baseline studies;
*   principal viewpoints to be covered;
*   system to be used for judging impact significance;
*   alternatives;
*   other developments if cumulative impacts are to be assessed.

3.5   An overall **description of the development** is needed. This should include a description of the siting, layout and *essential* characteristics of the development within its landscape context. The desirability of exploring alternatives and the time involved should be considered. These are discussed further in later sections.

3.6   The purpose of **baseline studies** is to describe, classify and evaluate the existing landscape and visual resource, focusing particularly on its sensitivity and ability to accommodate change. Baseline studies of landscape and visual resources play an important part not only in the EA process but also in the design process, providing an overview of the natural environmental constraints or opportunities which may influence the nature of the final development.

3.7   The critical stage of the assessment process is the systematic identification of **potential impacts,** prediction of their **magnitude,** and the assessment of their **significance**. Although impact assessment is primarily the responsibility of the developer and their advisers, views of the determining authority, relevant statutory consultees, conservation bodies and local residents should be taken into account. Often, agreement can be reached on many aspects of the assessment, hence focusing debate on only the contentious aspects, (see Consultation in Part 4).

3.8   The purpose of **mitigation** is to reduce as far as possible the potential environmental impacts of the development, employing strategies of avoidance, reduction, remediation and compensation. Mitigation is not solely concerned with 'damage limitation' but should also examine the potential for environmental enhancement through good scheme design as a separate objective. The need to monitor landscape and visual effects should be considered. This may be required where there is some uncertainty surrounding the predicted impact or the effectiveness of the mitigation measures or where a particularly sensitive landscape or visual resource may be affected.

3.9   EA is essentially an exercise in communication and the **presentation of findings** should include use of plain language and a clear structure, and good illustrative material. A concise non-technical summary will be an essential requirement as an aid to communication. The choice of a wide range of presentation techniques should be guided by the principles of value for money, simplicity, accuracy and suitability to the purpose in hand.

Description of the development

3.10   The value of a sound description of a proposed development is frequently underestimated. It can make an important contribution to the credibility and effectiveness of the study. To be prepared well, there needs to be good communication between members of the project team and the recognition that at some point the design must be 'frozen' to permit assessment to proceed upon firm assumptions, forming the factual basis for impact identification and prediction.

3.11   When preparing an ES, a general description of the development as a whole should be provided in the early sections as background to the subsequent technical assessment chapters. A more detailed description will then usually be needed within each of the assessment chapters. This description will highlight relevant sources of impact, and where possible will also give quantitative data.

3.12   A common issue in preparing project descriptions to accompany landscape and visual impact assessments, is how far to go in describing the proposed landscape design for the scheme. If good environmental planning and design principles have been applied, a high degree of mitigation will have been built into the scheme from the outset. Where possible, a distinction should be made between landscape measures designed specifically for mitigation and those which form mainstream components of the project design.

3.13   The description should focus upon factual explanation of the basic design elements, such as: access, layout, buildings, structures, ground modelling and planting, in so far as they affect landscape and visual resources. The design philosophy and the benefits of the design proposals are more suitably described in the section on mitigation, and even then should concentrate upon the way in which significant impacts have been addressed, rather than upon the merits of the design alone.

Consideration of alternatives

3.14   Within the general project description at the beginning of the study, project alternatives should be explored. Although not at present a formal requirement of the EA Regulations, this is likely to assume growing importance in future with moves toward sustainable development. This will require greater emphasis upon wider issues than the present simple focus upon site and project specific issues. Increasingly, the need for the development will have to be substantiated, and the overall strategy for siting, materials and energy use, transport and related issues will have to be explained.

3.15 At an immediate, practical level, there are several types of alternative that can and should be explored for every project.

These include different:
* locations, sites and alignments for linear developments;
* size/scale of development;
* site layouts, access and servicing arrangements;
* scheme design and processes;

and the 'do nothing situation', against which the development will be compared. The benefit of exploring alternatives is that they often offer the best scope for effective impact mitigation, especially in relation to landscape and visual impacts. For example, if there are serious landscape constraints associated with a particular site, the selection of an alternative location may be the only solution. Failure to recognise this at an early stage may result in a costly planning refusal. In the event that the developer wishes to proceed with the original proposal, the risks of doing so should be made explicit.

3.16 Depending on the type of study that is being prepared and the stage reached in the assessment process, more than one project alternative may be taken forward for comparative assessment, and hence more than one detailed project description will be required. The most common example is in the transport field where route option appraisals are frequently undertaken before a decision is made upon a preferred route for which a more detailed assessment will be carried out. Many other types of project would benefit from a similar approach.

Stages in the project life-cycle

3.17 In addition to consideration of alternatives, it should be recognised that project characteristics, and hence sources of impact, will vary through time. Construction, operation, decommissioning and restoration are characterised by quite different physical elements and activities. A separate, self-contained description of the development at each stage in the life-cycle will greatly assist landscape and visual impact prediction.

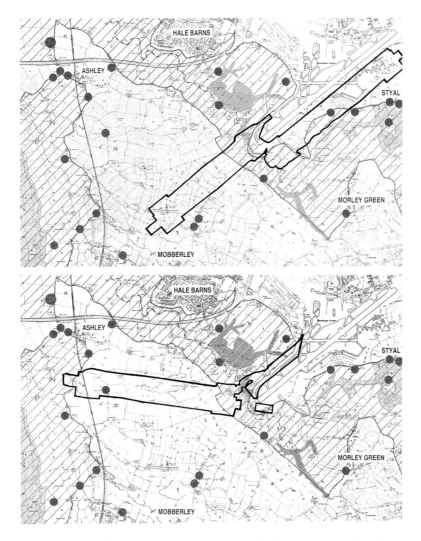

In this example illustrating alternative layouts, two options for runway alignments have been assessed.

3.18   For the **construction stage** of the project, the description should include where relevant:

- site access and haul routes, including traffic movements (which often differ from permanent access proposals);
- cut, fill, borrow and disposal areas;
- materials origins;
- materials stockpiles;
- staging areas;
- construction equipment and plant;
- utilities, including water, drainage, power and lighting;
- temporary parking and on-site accommodation and working areas;
- protection of existing features.

3.19   During the **operational stage,** the following matters will be most relevant to the landscape and visual impact assessment:

- access;
- infrastructure;
- buildings and other structures;
- delivery, loading and unloading areas;
- outdoor activities;
- materials storage;
- land management operations and objectives;
- utilities;
- workforce parking;
- landform and structure planting and hard landscape features;
- entrances, signs and boundary treatments;
- areas of possible future development.

3.20   In addition, **decommissioning and restoration** should be addressed. Important aspects here will include:

- access;
- after-use potential;
- residual buildings and structures;
- disposal or recycling of wastes and residues;
- restoration requirements, including materials and plant.

Information requirements

3.21   For each stage in the project life-cycle, similar types of data are needed to assist in landscape and visual impact predictions. Both **qualitative** and **quantitative** data are required, including for instance:

- form (including shape, bulk, pattern, edges, orientation, complexity and symmetry);
- materials (including texture, colour, shade, reflectivity, opacity);
- design (including layout, scale, style, distinctiveness);
- programme and duration of key site activities;
- site areas under different uses;
- physical dimensions of major plant, buildings and structures;
- volumes of material;
- numbers of scheme components such as houses and parking spaces;
- movements of plant, materials and workforce.

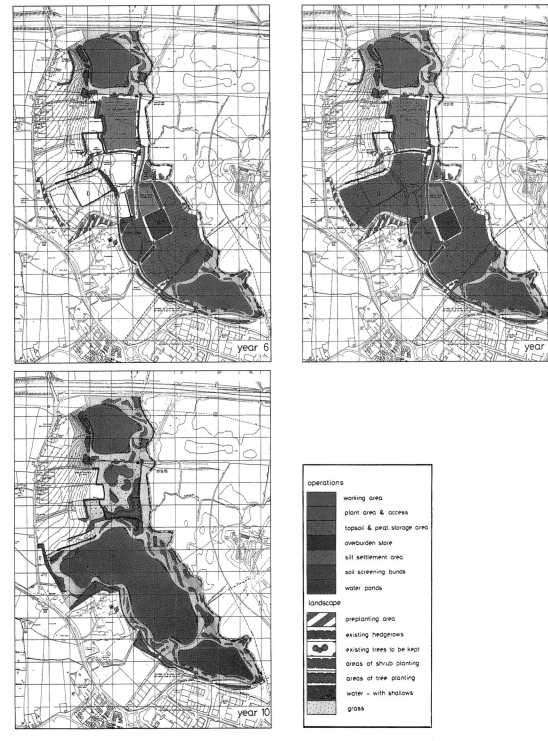

operations

working area

plant area & access

topsoil & peat. storage area

overburden store

silt settlement area

soil screening bunds

water ponds

landscape

preplanting area

existing hedgerows

existing trees to be kept

areas of shrub planting

areas of tree planting

water - with shallows

grass

Using clear and concise graphic techniques this series of drawings communicates the sequence of parallel operations where minerals are extracted and a new landscape is formed

3.22   It is recognised that it is often difficult to provide accurate and complete information on all these varied aspects of a development. Nonetheless, the importance of such information cannot be over stressed, as it is the foundation for all impact predictions. Where key data on project characteristics are lacking, there may be a need to make explicit assumptions as to what will happen, based upon the 'worst case situation'.

Off-site and indirect impacts

3.23   Finally, neither the project description nor subsequent stages of the assessment should neglect off-site and indirect impacts. For example, alterations to the drainage regime in the vicinity of the site could result in changes to the vegetative cover and a consequent change to the landscape character, such impacts should be assessed. Off-site impacts may also occur as a result of *associated development*, for example, a transmission line to a power station. This is a somewhat grey area in EA procedures because on the one hand the EC Directive [1] implies that indirect effects are relevant and should be considered; while on the other hand any associated developments outside the site boundary are usually subject to separate consent procedures.

Landscape and visual impacts can occur away from the site. In this example, spoil from a large construction project is deposited at some distance from the site

3.24   These guidelines recommend that it is good practice to at least draw attention to off-site and indirect impacts resulting from *associated development* and indicating, where relevant, that this should constitute a separate field of study. Common off-site matters to be addressed include:

- upgrading of transport infrastructure and new signs;
- mineral extraction and waste disposal requirements;
- new or improved off-site utilities such as water and waste water treatment plants, surface water drainage systems, gas pipelines, electricity substations and transmission lines, and telecommunications facilities;
- transport implications;
- cumulative impacts in association with other separate developments.

3.25   Indirect impacts of development should also be acknowledged. Examples include pressure for housing development associated with a large new industry; retail development in response to a new road junction; and increased recreational activity following improvements to access. These issues should be dealt with primarily by the planning authority as part of the development planning process, but in practical terms should not be ignored in the study. In addition these issues will almost certainly be raised by consultees and the public.

The construction of a new road junction has lead indirectly to the development of retail units and their associated signage

Baseline studies

3.26   The initial step in any landscape and visual impact assessment is to review the existing landscape and visual resource: this process results in a 'baseline report'. The data collected will form the basis against which to review the magnitude and the significance of the predicted landscape and visual impacts of the development. It is important to bear in mind that baseline conditions are not static. The landscape may already be changing for reasons unrelated to the development itself. Hence the baseline studies should look not only at existing landscape conditions, but at the landscape dynamic, and should take account of any landscape management strategy or guidelines that exist or are in preparation.

Purpose and scope

3.27   The purpose of baseline studies is to record and analyse the existing **character, quality, enhancement potential and sensitivity** of the landscape and visual resources in the vicinity of the proposed development. This will require description, classification and evaluation of these resources:

- **Description** is the process of collecting and presenting information about landscape and visual resources in a systematic manner.

- **Classification** is the more analytical activity whereby landscape resources, in particular, are sorted into units of distinct and recognisable character.

- **Evaluation** means attaching a (non-monetary) value to a given landscape or visual resource, by reference to specified criteria.

3.28   The level of detail should be appropriate to the scale of the development, to the sensitivity of the receptors and to the potential for impacts to occur. It should also be appropriate to the early stages of the assessment process. For instance, at the scoping stage, the primary aim is to identify key issues and constraints. For this purpose, fairly broad-brush, mainly desk-based studies may be adequate, for example recording landscape designations, areas of ancient woodland, and known sites of recreational interest that will influence site selection. Statutory conservation agencies and local planning authorities are likely to be the main sources of information for this.

By the time an application for consent and an ES for a specific development are ready to be submitted, much more detailed baseline studies should have been carried out as part of a full impact assessment.

3.29   The baseline studies for most landscape and visual impact assessments will need to include three stages: **desk study, field survey** and **analysis.** The paragraphs that follow describe the principal information sources and techniques that are appropriate at each of these stages. Further information on baseline landscape assessment can be found in advice from the Countryside Commission [7], Scottish Natural Heritage [11] and the Countryside Council for Wales [12].

Desk study

3.30   When commencing a landscape and visual impact assessment, it is essential at the outset to briefly visit the site and review existing map and written data about the development site or area. These studies initially will need to extend well beyond the development site, both to assist in site selection, and to help establish the wider landscape setting and context. As a minimum, the baseline studies should extend to cover the whole of the area from which the development is visible and usually the wider landscape context.

3.31   Useful sources may include:

- current and historical Ordnance Survey and other maps;
- geology, soils and land use maps, hydrological survey;
- aerial photographs;
- structure and local plans showing landscape designations and other relevant planning policies (including associated survey and issues reports);
- informal planning documents such as countryside strategies and landscape assessments or guidelines;
- data on archaeology, ecology and buildings and settlements and other conservation interests within the area;
- common land and Rights of Way maps;
- Meteorological Office data.

3.32   Information of relevance to the proposed development should be extracted and summarised from these sources, either as written text or in map form. In particular, the desk study should explore:

- **patterns and scale of landform, landcover and built develop-**

ment, which give guidance on the general landscape character of the surrounding area;

- **any special values** that may apply, notably national landscape designations (such as National Parks, Heritage Coasts, AONBs, and NSAs), local authority landscape designations (AGLV, SLA) and in an urban context, Conservation Areas, and any historical and cultural associations;

- **specific potential receptors of landscape and visual impact**, which may include important components of the landscape, as well as residents, visitors, travellers through the area and other groups of viewers.

Field survey

3.33   The desk study should provide a sound basis for subsequent field survey work. For instance, it may define draft landscape character areas around the development, delineate the likely zone of visual influence, identify the principal viewpoints, and highlight sensitive receptors. All these findings will require confirmation in the field, particularly in an urban and urban fringe setting, where map and even air photograph data may well be out-of-date. Ideally, the field survey should be carried out by more than one person so that discussion can take place and a consensus view obtained between professionals as part of the process of verification.

3.34   In relation to **landscape character,** existing guidance from the Countryside Commission [7] gives detailed advice on techniques for field survey and assessment. The most commonly used technique involves the completion of a structured field survey form for selected viewpoints across the study area. The survey form permits recording of both objective description and subjective impressions of the landscape, as well as details of landscape condition and trends for change. After completion of the field work, a classification of the landscape into units of common character can be prepared through a process of analysis.

3.35   The approximate **visibility** of the development should be determined through topographic analysis, either manually or by computer. The actual extent of visibility always needs checking in the field because of the localised screening effects of buildings, walls, fences, trees, hedgerows and banks. Knowledge of the precise siting and dimensions of the proposed development is also critical, and this may mean that visibility surveys need to be repeated as siting, layout

FIELD SURVEY FORM

Viewpoint No.:   **Location:**   **Date:**

Film/Photo. Nos.:   **Direction of view:**

Annotated sketch

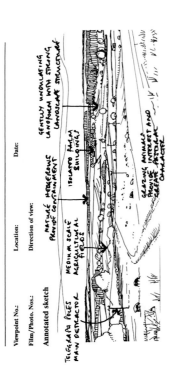

Brief description (describe the main elements and features and the way in which they are organised also note any special aesthetic factors including detractors and attractors)

[handwritten notes - brief description]

Landcover and landscape elements

Note the dominant elements in the landscape

farm buildings	walls	deciduous
churches	fences	woodland (type)
masts, poles	hedges	plantation
pylons	banks	isolated trees
industry	shelterbelt	tree clumps
vernacular buildings	field pattern	hedgerow trees
settlement (type)	arable	parkland
built-up	pasture	scrub
mineral working	orchards	marsh
ruins		

river	footpath
stream	track
lake	road
reservoir	motorway
pond	railway
canal	
waterfall	
beach	
dune	
mudflat	

Landform

flat	plain
rolling	rolling lowland
steep	plateau
vertical	hills
	scarp
	cliff

coast
estuary
broad valley
narrow valley
deep gorge

Aesthetic factors

BALANCE:	balanced	harmonious	discordant
SCALE:	small	intimate	medium	large
ENCLOSURE:	enclosed	confined	open	exposed
TEXTURE:	textured	smooth	rough	very rough
COLOUR:	muted	monochrome	colourful	garish
DIVERSITY:	simple	uniform	diverse	complex
MOVEMENT:	vacant	remote	peaceful	active
UNITY:	interrupted	unified	fragmented	chaotic
FORM:	angular	straight	curved	sinuous
SECURITY:	safe	comfortable	unsettling	threatening
STIMULUS:	boring	bland	interesting	invigorating
PLEASURE:	offensive	unpleasant	pleasant	beautiful

Landscape Condition

An intensively managed agricultural landscape with well maintained trees and agricultural buildings. Noticeable lack of detracting features w urban/fringe influences. The overall condition is good.

Most Appropriate Management Strategy

Conservation

[handwritten notes]

Restoration

Reconstruction

Ability to Accommodate Change

[handwritten notes about agricultural landscape and development]

A typical field survey form for landscape assessment, allowing both factual information and subjective impressions to be recorded

and design proposals are progressively refined.

3.36   **Principal viewpoints** within the study area should also be identified. Both public and private viewpoints are relevant. As in the survey of landscape character, both objective description and subjective impressions should be recorded. Use of a field survey form such as that recommended by the Countryside Commission [7] may again be helpful and the visual survey should be supported by a comprehensive photographic record from principal viewpoints.

3.37   Lastly, the field survey should identify and address specific **sensitive receptors.** These may include landscape elements and features that may be directly affected by the development, as well as residents, visitors and other groups of viewers. In the case of landscape receptors, the field survey should record topographic, geological and drainage features; woodland, tree and hedgerow cover; land use, field boundaries and artifacts; archaeological, historic and cultural features; access and rights of way. In the case of visual receptors the types of viewers affected, an estimate of their numbers, or whether they are few or many, duration of viewing, and potential seasonal screening effects should be noted.

Analysis

3.38   On completion of the desk study and field survey work, the results should be analysed and written up. The findings should be presented in a clear, structured fashion, as they will be a key component of the landscape and visual impact assessment as a whole. However, a clear distinction should be made between first, the description and assessment of the individual elements or features of a landscape and their importance, and secondly, the synthesised character of distinctive areas. The findings will also play a crucial role in guiding scheme design, and as such they should be useful to the whole design team for the project, which may include engineers, architects and other professionals. Further guidance on presentation is given in chapter 8. However, the baseline studies section of the report should cover those aspects of the landscape and visual resource set out below.

3.39   **Scale and character.** A concise description of the existing character of surrounding landscapes should be given. This should review, briefly, the physical and human influences that have helped to shape the landscape and any current trends for change. It should identify the landscape's distinctive elements, features and their spatial organisation, and should be clearly illustrated by photographs or

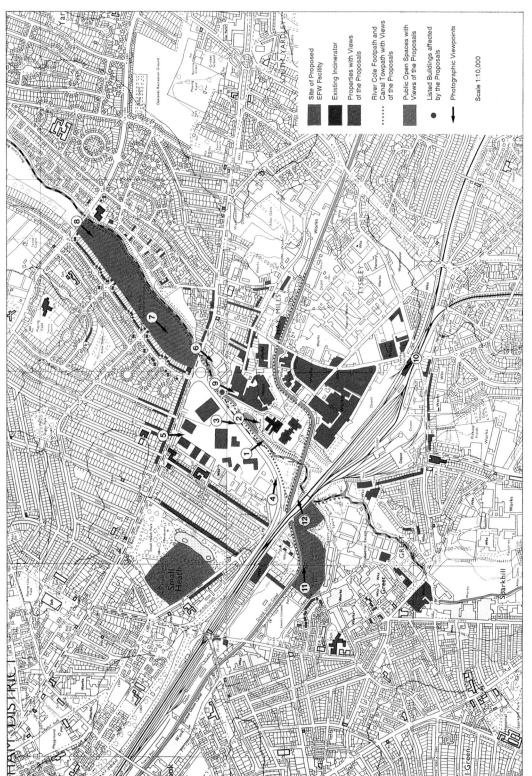

The location of principal viewpoints, properties with views of the development, public open spaces with views and the development site itself, are clearly illustrated in this map

LANDSCAPE TYPES

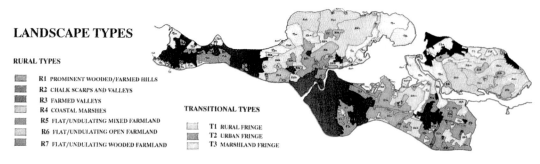

RURAL TYPES

- R1 PROMINENT WOODED/FARMED HILLS
- R2 CHALK SCARPS AND VALLEYS
- R3 FARMED VALLEYS
- R4 COASTAL MARSHES
- R5 FLAT/UNDULATING MIXED FARMLAND
- R6 FLAT/UNDULATING OPEN FARMLAND
- R7 FLAT/UNDULATING WOODED FARMLAND

TRANSITIONAL TYPES

- T1 RURAL FRINGE
- T2 URBAN FRINGE
- T3 MARSHLAND FRINGE

LANDSCAPE CHARACTER AREAS

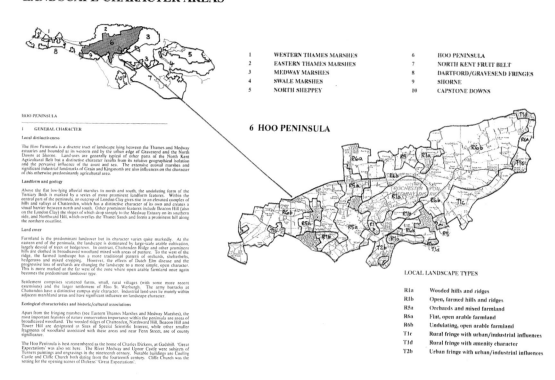

1	WESTERN THAMES MARSHES	6	HOO PENINSULA
2	EASTERN THAMES MARSHES	7	NORTH KENT FRUIT BELT
3	MEDWAY MARSHES	8	DARTFORD/GRAVESEND FRINGES
4	SWALE MARSHES	9	SHORNE
5	NORTH SHEPPEY	10	CAPSTONE DOWNS

HOO PENINSULA

1   GENERAL CHARACTER

Local distinctiveness

The Hoo Peninsula is a discrete tract of landscape lying between the Thames and Medway estuaries and bounded at its western end by the urban edge of Gravesend and the North Downs at Shorne. Land-uses are generally typical of other parts of the North Kent Agricultural Belt but a distinctive character results from its relative geographical isolation and the pervasive influence of the coast and sea. The extensive coastal marshes and significant industrial landmarks of Grain and Kingsnorth are also influences on the character of this otherwise predominantly agricultural area.

Landform and geology

Above the flat low-lying alluvial marshes to north and south, the undulating form of the Tertiary Beds is marked by a series of more prominent landform features. Within the central part of the peninsula, an outcrop of London Clay gives rise to an elevated complex of hills and valleys at Chattenden, which has a distinctive character of its own and creates a visual barrier between north and south. Other prominent features include Beacon Hill (also on the London Clay) the slopes of which drop steeply to the Medway Estuary on its southern side, and Northward Hill, which overlies the Thanet Sands and forms a prominent hill along the northern coastline.

Land cover

Farmland is the predominant landcover but its character varies quite markedly. At the eastern end of the peninsula, the landscape is dominated by large-scale arable cultivation, largely devoid of trees or hedgerows. In contrast, Chattenden Ridge and other prominent hills are clothed in broadleaved woodland mixed with areas of pasture. To the west of the ridge, the farmed landscape has a more traditional pattern of orchards, shelterbelts, hedgerows and mixed cropping. However, the effects of Dutch Elm disease and the progressive loss of orchards are changing the landscape to a more simple, open character. This is more marked at the far west of the zone where open arable farmland once again becomes the predominant landcover type.

Settlement comprises scattered farms, small, rural villages (with some more recent extensions) and the larger settlement of Hoo St. Werburgh. The army barracks at Chattenden have a distinctive campus style character. Industrial land-uses lie mainly within adjacent marshland areas and have significant influence on landscape character.

Ecological characteristics and historic/cultural associations

Apart from the fringing marshes (see Eastern Thames Marshes and Medway Marshes), the most important features of nature conservation importance within the peninsula are areas of broadleaved woodland. The wooded ridges of Chattenden, Northward Hill, Beacon Hill and Tower Hill are designated as Sites of Special Scientific Interest, while other smaller fragments of woodland associated with these areas and near Fenn Street, are of county significance.

The Hoo Peninsula is best remembered as the home of Charles Dickens, at Gadshill. 'Great Expectations' was also set here. The River Medway and Upnor Castle were subjects of Turners paintings and engravings in the nineteenth century. Notable buildings are Cooling Castle and Cliffe Church both dating from the fourteenth century. Cliffe Church was the setting for the opening scenes of Dickens' 'Great Expectations'.

6   HOO PENINSULA

LOCAL LANDSCAPE TYPES

R1a	Wooded hills and ridges
R1b	Open, farmed hills and ridges
R5a	Orchards and mixed farmland
R6a	Flat, open arable farmland
R6b	Undulating, open arable farmland
T1c	Rural fringe with urban/industrial influences
T1d	Rural fringe with amenity character
T2b	Urban fringe with urban/industrial influences

LANDSCAPE ENHANCEMENT STRATEGY

ENHANCEMENT STRATEGY

- CONSERVATION
- RESTORATION: REPAIR
- RESTORATION: REINSTATE
- RECONSTRUCTION

Example of the process of landscape characterisation and allocation of appropriate enhancement strategies

3.40   Often the character description will include a classification of the landscape into distinctive character areas or types, usually based on variations in landform and landcover. This classification should take into account other landscape assessments that may have been prepared previously for the study area, for example by the county or district council. It should reflect all aspects of landscape character, from the scenic or visual through to archaeology, history, ecology, built environment and cultural associations.

3.41   **Condition and importance.** Qualitative analysis requires an assessment to be made of landscape **condition** as well as a judgement regarding its **importance** in the sense of aesthetic or cultural value. The analysis should draw upon both desk study and field survey work and should be fully supported by both illustrations and documentary evidence, which may include:

* a list of any landscape designations that may apply;
* a summary of the reasons for any landscape designation, for example where a landscape type is rare in a national or regional context;
* professional judgements as to the scenic quality of the site and its wider landscape context and to the importance of landscape components;
* assessment of the condition of the important landscape components, including management of land, vegetation, buildings and other features, and the extent of deviation from the perceived optimum condition;
* details of any other notable conservation interests such as features of historical, wildlife or architectural importance;
* reference to any special cultural associations, such as important writings and paintings that feature local landscapes;
* past and present perceptions of local value.

3.42   The Countryside Commission's landscape assessment guidance [7] gives further advice on criteria for evaluating landscape quality in England.

3.43   **Sensitivity.** At the same time, conclusions should be drawn as to the overall sensitivity of the landscape and visual environment to the type of development envisaged. Although, at the baseline studies stage, the detailed characteristics of the development may still not be known, the broad nature and extent of potential landscape and visual impacts should be clear. By analysing the general

Landscape and visual
constraints map

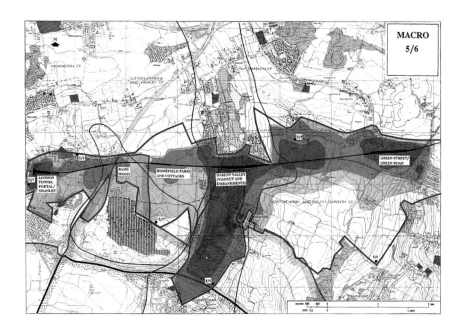

MACRO SURVEY

Visual And Landscape Impact

Severe

Moderate

Low

**All Other Areas Within The Visual Envelope
Will Suffer Minimal Landscape And Visual Impact**

CTRL Sources Of Impact

Severe

Moderate   ━┫   Tunnel Portal

Low

– ‧ – ‧   CTRL Tunnel

Visual Envelope

Character Area

●   Viewpoint

LANDSCAPE SURVEY

Woodland

Recreational Area

Major Water Feature

Area of Outstanding Natural Beauty

Special Landscape Area

Historic Park/Garden

Conservation Area

Site of Nature Conservation Interest

Area of High Nature Conservation Value

Site of Special Scientific Interest

Major Footpath

Area Subject to Tree Preservation Order

●   Tree Subject to Tree Preservation Order

pattern of landscape character and quality, the detailed distribution of landscape and visual receptors, and the extent to which these factors will be tolerant of change, the principal sensitivities and constraints upon development will become apparent. Often these constraints can be summarised in map or plan form.

3.44   **Change or enhancement potential.**   Analysis of character, condition, aesthetic quality and sensitivity to change should together provide pointers to the potential and desirability for landscape enhancement. It should, for example, be possible to identify:

- those landscapes which exhibit a strong character and sense of place or have many features that are notable, for example, their scenic, historical or ecological interest;

- those landscapes where individual elements or features have suffered decline, but where there is still scope to restore the typical character, or aspects of it;

- those landscapes where the overall character has been significantly altered, so that reconstruction, or even creation of a new landscape is required.

3.45   These are not rigid distinctions and clearly form part of a continuum of landscape quality and condition. However such an analysis can give general guidance as to where, and how, new development can be sensitively accommodated in the landscape. It can also indicate important issues of degradation or intrusion which it may be possible to tackle through landscape or environmental enhancement or planning gain associated with new development.

3.46   **Visual analysis.**   Following the field survey, the extent to which the existing site is visible from surrounding areas can be confirmed and presented on a plan, identifying specific elements, such as landform, buildings or vegetation, which may interrupt, filter or otherwise influence views. The locations of principal viewpoints should also be mapped and these existing views illustrated by photographs or sketches or both with annotations which describe any important characteristics that might be of relevance to the assessment of impact or design of mitigation.

3.47   By the end of the baseline studies, it should be possible to advise, in landscape and visual terms, on the development's acceptability in principle and its preferred siting, layout and design. This information should be communicated to the developer and to all members of the project design team to contribute to influencing scheme design.

Impact assessment

3.48   This stage in the process aims to:

*   identify systematically all the potential landscape and visual impacts of the development;

*   predict and estimate their magnitude as accurately as possible;

*   assess their significance in a logical and well-reasoned fashion.

3.49   The assessment should describe the changes in the character and quality of the landscape and visual resources that are expected to result from the development. It should cover both landscape impacts, that is changes in the fabric and character of the landscape; and visual impacts, that is changes in available views of the landscape and the effect of those changes on people.

Impact identification

3.50   The first task is the systematic comparison of sources of impact with landscape and visual receptors. Checklists and matrices can often assist in this process, for example, a matrix which shows on one axis different construction activities and built elements, and on the other axis the principal landscape and visual receptors of the site, may assist in the initial identification of key impacts for further study.

3.51   It is important at this stage to consider potential impacts at different stages in the life-cycle of the development. Hence construction stage impacts should be examined separately from operational stage impacts. The longer term impacts of decommissioning the site or facility should also be borne in mind. For instance, in relation to power stations, the tallest site structures may be the construction cranes, whose impact is sometimes ignored because it is considered to be temporary, even though construction may last for several years or more. Similarly, decommissioning of a power plant may have significant impacts, not only in terms of construction activity, but also in the form of residual structures such as reactor cores, which may need to remain for many years after the plant has ceased to operate.

3.52 **Landscape impacts.** These include the direct and indirect impacts of the development upon landscape elements and features as well as the effect upon the general landscape character and quality of the surrounding area. For example, the main direct impact of a flood alleviation scheme may be a loss of vegetation due to embankment raising and strengthening. However, there may also be indirect, long term impacts such as the drying of a wet meadow following changes in the hydrological regime.

3.53 Direct impacts can be shown on a plan and can also be described clearly and objectively. Where possible the description should quantify the extent and duration of any damage or loss. For example, it may state how many mature trees and how many metres of hedgerow are to be lost as a result of landfill development; how long the workings will be active; and how much new planting will take place at restoration. This type of factual data will be especially useful when comparing different scheme options and has the advantage of helping to put in context the degree of change that will occur.

3.54 Wider impacts upon landscape character and quality are less easy to predict objectively. Interpretation and professional judgement will need to be applied. A clear picture of likely impacts should be presented by referring back to the baseline landscape character assessment and describing - using maps, illustrations and plain language - how the development may alter existing patterns of landscape elements and features. Where indirect impacts are considered

	CONSTRUCTION										OPERATION AND MAINTENANCE										
	Primary Access Road Construction	Site Clearance (initial/continuing)	Excavation (initial/continuing)	Other Tunnelling/Mining/Quarrying Techniques	Drainage Alteration	Groundwater Manipulation (drawdown/flow modification)	General Building/Infrastructure Construction	Waste Management Unit (treatment/disposal) Construction	Equipment Operation (day)	Utilities (transmission lines/pipelines) Installation	Forest/Vegetative Clearance (initial/continuing)	Excavation (initial/continuing)	Tunnelling/Mining/Quarrying (initial/continuing)	Spoil and Overburden Storage	Waste Import (eg landfill, by road)	Material Export (eg minerals extraction, by road)	Compaction (eg landfill)	Groundwater Manipulation (drawdown/flow modification)	Operational Emissions to the Atmosphere	Operational Emissions to Surface Waters	Operational Emissions to Groundwater
LANDSCAPE RECEPTORS																					
East Wolds AONB																					
Beacon Hill Beauty Spot																					
Oak Wood (along western boundary of site)																					
Hedgerows																					
VISUAL RECEPTORS																					
'East View Terrace'																					
'Copse Avenue'																					
'Whitegates House'																					
Kirk Valley Country Park																					
'Green Lane Footpath'																					
M6 (Southbound)																					
Grange to Leyland Road																					

Landscape and visual impact identification matrix: Matrices can be a useful tool for assisting in the identification of landscape and visual impacts, and in presenting large amounts of information in a concise form

48

to be beyond the scope of the assessment, for example when a new road scheme is likely to be followed by housing or industrial development, the location and extent of which is still unknown, this limitation should be acknowledged.

3.55 **Visual impacts.** In predicting visual impacts, the main requirements are to show:

- the extent of potential/theoretical visibility;
- the views and viewers affected;
- the degree of visual intrusion or obstruction that will occur;
- the distance of view;
- the resultant impacts upon the character and quality of views.

3.56 Numbers and types of viewers affected can be quantified and map symbols or tones used to denote the distribution of major and minor visual impacts. The degree of visual intrusion should be described by reference to building mass, height and colour, and reference should also be made to the different degrees of screening that will apply in summer and winter.

3.57 The qualitative effects of the development on the character and quality of views will need to be addressed with care. Photomontages, video presentations and other forms of visualisation should supplement written description where appropriate. The selection of viewpoints is critical and should closely reflect the principal viewpoints identified during the baseline studies. In addition, the 'worst case' principle should be applied, that is, winter views from the most severely affected viewpoints should be chosen. It is often useful to agree viewpoints in advance with the planning or other determining authority that will be the recipient of the study.

Assessing significance

3.58 In assessing the significance of landscape and visual impacts, reliance should be placed upon commonsense and reasoned judgement, supported wherever possible by substantiated evidence. Conclusions should be based upon a combination of factors, including:

- **The sensitivity of the affected landscape and visual resources.** Here the development context and the character, importance, condition and tolerance of change of the existing landscape are relevant. For instance, within an industrial landscape with little historical or other interest, a new sewage treatment works

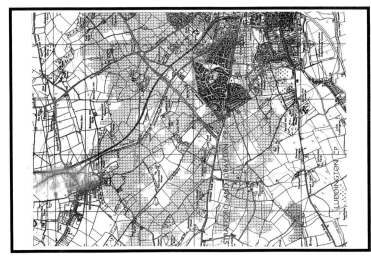

VISUAL ENVELOPE MAP

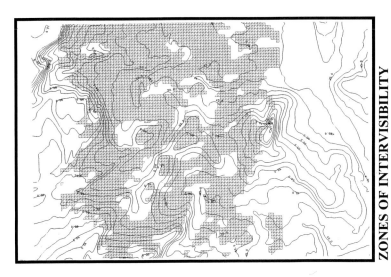

ZONES OF INTERVISIBILITY

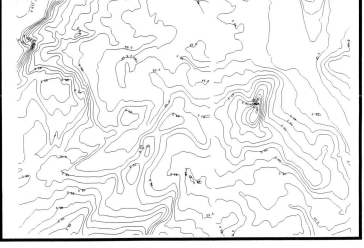

CONTOUR PLAN

The extent of the theoretical visibility can be shown by a Zone of Visual Influence (ZVI). In this series of illustrations a computer is used to generate a ZVI, or in this case a Visual Envelope Map (VEM).

may have an insignificant landscape and visual impact; whereas in a *valued* landscape it might be considered to be an intrusive element. Important components are inherently more sensitive when combined with a low tolerance to change. In such a case, even changes of relatively low magnitude may be considered a highly significant impact.

- **Impact magnitude.** Considerations such as impact magnitude and duration are very important in determining the significance of impacts. For example, a temporary change that is confined to a small area and visible only from a few private dwellings, may be considered to be of low magnitude. However, it would assume greater significance if it is permanent and cannot be mitigated.

- **Whether impacts are beneficial or adverse.** There will be a need to take a view as to the overall impact of the development on landscape character and quality, and the specific impact on the various sensitive receptors. The impact on certain receptors may be beneficial, while that on others may be detrimental.

- **Professional judgement.** Landscape professionals are trained and experienced in advising upon the relationship between a new development and its surroundings. This enables them to make a reasoned judgement on the degree of damage and visual intrusion that development may bring.

- **Views expressed during consultations.** The opinions of local residents and representative bodies should always be sought and considered. Often consultations may reveal certain areas or features of the local landscape that have special amenity value for local people or an unforeseen conservation interest.

3.59   The relative weight given to each of these various factors, and the rationale behind the conclusions that are drawn, should be made explicit within the landscape and visual impact assessment. In presenting the assessment findings, it may be useful to include separate statements of impact significance both before and after mitigation. This helps to show what has been achieved during the design process and gives an indication of the effectiveness of the mitigation measures.

3.60   Some of the proposed mitigation measures, especially those involving planting, may not be as successful as predicted, may fail altogether, or take a long time to become established. This rein-

Examples of photomontages showing the degree of visual intrusion

forces the importance of assessing the impacts of the development **before** mitigation as well as with these measures in place.

Criteria and thresholds

3.61   The development of thresholds of impact significance is recommended as a way to standardise the conclusions of the landscape and visual impact assessment so that they are consistent within themselves, and permit the comparison of different project options, and different categories of impact.

3.62   **Criteria.** In order to develop significance thresholds it is necessary first to classify the sensitivity of receptors and the magnitude of change according to reference points along a continuum, as in Figure 3.1 below. In this case scale of 'high, medium and low' has been used, but it must be stressed that this is only an example. Every project will require its own set of criteria and thresholds, tailored to suit local conditions and circumstances, and it should be remembered that impacts can be positive as well as negative. The benefit of such a system, though, is to help separate fact from interpretation, and hence to simplify discussion and agreement on the significance of impacts.

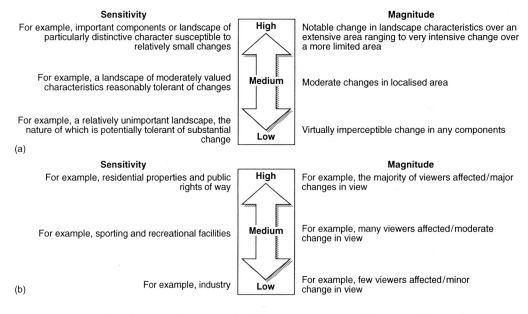

Figure 3.1: (a) Classification of sensitive landscape receptors and impact magnitude.
(b) Classification of sensitive visual receptors and impact magnitude.

3.63 **Thresholds.** Significance thresholds can be determined from different combinations of sensitivity and magnitude.

- Example 1:'**Substantial' impacts** can be a product of high sensitivity or high magnitude
- Example 2: **'Moderate' impacts** can result from medium sensitivity and magnitude, or low sensitivity with high magnitude
- Example 3: **'Slight' impacts** can be a product of low sensitivity or low magnitude

Some examples are shown in figure 3.2

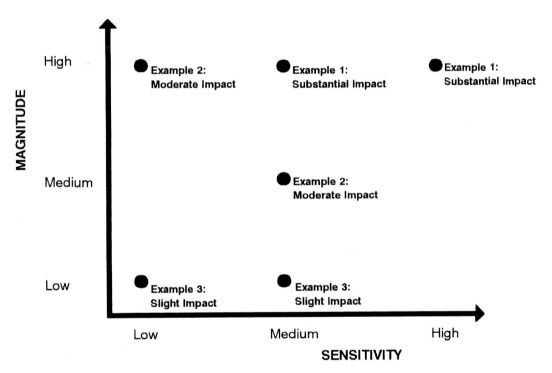

Figure 3.2: The relationship between sensitivity and magnitude in defining significance thresholds

3.64 Numerical scoring or weighting should be avoided. Attempting to attach precise numerical values to qualitative resources is rarely successful, and should not be used as a substitute for reasoned professional judgement.

Mitigation

3.65   These guidelines use the term 'mitigation' because it is now common currency in describing measures to deal with environmental impact. Mitigation should not be an afterthought, nor something that is applied to the final scheme design to soften its more obvious adverse effects. If this approach is adopted, mitigation only serves to mask what would otherwise be an unacceptable design, rather than dealing with the underlying problems. Mitigation should be appropriate, adequate and enforceable in the long term. Developers should demonstrate that long term control is secured, for example, where mitigation is on someone else's land.

3.66   The section on mitigation is frequently found towards the end of the ES. Although this is a logical and convenient place in which to summarise the mitigation measures, it does not emphasise sufficiently the true relevance of mitigation to the project planning and design process. Mitigation is a design skill which should start at the very inception of a project with the analysis of environmental opportunities and constraints. It should be used to adapt and modify the development to take account of these factors, and hence achieve the best environmental fit. At every stage in landscape and visual impact assessment, there should be feedback to project planning and design, to achieve the minimum adverse impact and optimum benefit for landscape character and visual amenity.

3.67   Environmental enhancement should be a primary objective for every landscape and visual impact assessment. Enhancement is closely linked to mitigation, but explores the scope for a development project to contribute positively to the landscape of the development site and its wider setting. Therefore enhancement proposals should be based on a sound initial assessment of landscape character, quality and trends for change. Can the development help restore or reconstruct local landscape character and local distinctiveness? Can it assist in meeting local authority objectives for the area? Can it help solve specific issues such as derelict land reclamation? Enhancement may take many forms, including better management, or restoration of historic landscapes, habitats and other valued features; enrichment of denuded agricultural landscapes; measures to conserve and improve the attractiveness of town centres; and creation of new landscape, habitat and recreational areas. In this way environmental enhancement can make a very real contribution to sustainable development.

3.68   The three rules for successful mitigation are that it should be **effective, appropriate** and **feasible.** The application of the following simple principles can greatly increase the effectiveness of proposed mitigation:

• **Mitigation measures should** normally **be designed to suit the existing landscape character** and needs of the locality. They should respect and build upon landscape distinctiveness and help meet the main issues already facing the landscape.

• **All significant adverse impacts should be considered for mitigation,** and proposed mitigation measures should address specific issues and impacts. Objectives or performance standards should be set, describing exactly what is required for mitigation to be effective.

• **The developer should demonstrate a commitment to the implementation of mitigation measures** to an agreed programme and timetable. The ES should clearly set out which mitigation measures will be the developers' responsibility, and which might be undertaken by others.

• **Care should be taken to mitigate significant impacts occurring at all stages in the project life-cycle.** Mitigation proposals should cover not only the operational stage, but also construction, decommissioning and restoration.

• **It should be recognised that many mitigation measures, especially planting, are not immediately effective.** This implies a need to predict residual impacts for different periods of time, such as day of opening, year 5 and year 15.

• **Performance criteria should be proposed** by the developer for the management, maintenance and monitoring of new landscape features. Contingency plans should be clearly stated at the outset, in the event that mitigation measures are proved unsuccessful.

• **A programme of monitoring, appropriate to the development, should be agreed so that compliance and effectiveness can be easily monitored** and evaluated by the determining authority.

• The local community and interest groups should, as far as possible, be consulted over the proposed mitigation measures.

Strategies

3.69   The mitigation of impacts can take place at a number of levels. The *ideal* strategy for each identifiable impact is one of avoidance. If this is not possible, alternative strategies of reduction, remediation and compensation may each be explored. It is tempting for developers to think that landscape issues can be left to the later stages of scheme design. To do so nearly always results in increased mitigation costs, where early opportunities for avoidance of impact mitigation are missed. Remediation and compensation are generally more expensive than avoidance. Some of the main issues associated with different mitigation strategies are outlined below.

3.70   **Avoidance of impact.** Avoidance of impact through careful siting, planning and design should first be considered. The route of a major highway, for example, is usually sufficiently flexible to avoid the most important landscape constraints. Where the need for the road is great, and the environmental constraints intense, highway design standards may need to be adapted to suit the environmental conditions. For housing, commercial and industrial developments, the siting of buildings in relation to existing landform and vegetation is critical, and visual analysis, can greatly assist in arriving at the best solution. For almost every major development, much time, money and public concern can be saved if serious environmental constraints can be identified and avoided before design proposals are well advanced.

3.71   **Reduction of impact.** After considering impact avoidance, the reduction of any remaining conflict with environmental constraints is the next priority, which means attention to layout and to site levels. Setting a development into the ground can often help it to be absorbed into the landscape. It can also screen low level visual 'clutter' such as parking, outdoor storage and working areas. By using landform and ground modelling to screen the lower portion of tall structures, the human eye is much less able to judge their size. New woodland or other planting, provided it is of sufficient size, will often reinforce the beneficial effects of good site layout and will assist integration with the landform. However, it should be remembered that new landscape features may be visually intrusive if poorly designed. Stark, geometric earth bunding, often used for noise amelioration next to highways, is an example of an unsympathetic landscape treatment that could be improved with more space and

STAND TYPE: Mix E

STAND COMPOSITION

Climax spp: ash, beech
Sub-climax: birch, cherry, whitebeam
Understorey spp.:   field maple, dogwood, hazel, privet
Shrub Edge spp.:   assorted native trees and shrubs

MANAGEMENT OBJECTIVE

To create a two to three storey high forest which will help to screen the works area. The upper canopy will be of ash and beech, in which the predominant long term climax species will be beech. Occasional sub-climax trees (cherry and whitebeam) and a coppiced understorey are to be retained. The stand is to be bordered by a mixed tree/shrub edge. Manage the beech on a rotation of about 150 years.

MANAGEMENT POLICY

Thin to favour the climax and understorey species. Provide conditions suitable for the retention of occasional sub-climax (cherry and whitebeam) and understorey species. Within the climax blocks, remove the weak and supressed trees first. Coppice the understorey species as necessary within the thinning regime to release the climax species. Prevent growth of tall trees in the vicinity of power cables.

MANAGEMENT PRESCRIPTIONS

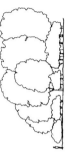

YEARS 1-5	**YEARS 5-10**	**YEARS 10-15**	**YEARS 15-20**	**YEARS 20-25**	**YEAR 25+**
Weed control and replacement of failures.	Thin edge shrubs.	-	Thin main body of planting to favour climax and understorey spp. Thin shrub edge.	Thin main body of planting to favour climax and understorey spp. Thin shrub edge.	Review every five years and thin as necessary.

Performance criteria for new planting can usefully be expressed in the form of management plans or maintenance schedules

gentle grading.

3.72   **Remediation of impact.** Mitigation measures that rely solely on what is commonly termed 'landscaping' to remedy the impacts of an otherwise fixed scheme, are unlikely to succeed. Remediation should be seen as part of the overall process of avoiding and reducing adverse impacts. Nevertheless, the sympathetic treatment of external areas, particularly where this employs substantial tree planting, will soften the effects of buildings in most circumstances and may bring about a general improvement in the local environment.

3.73   **Compensation for impact.** Serious, unavoidable impacts can sometimes be offset by related environmental improvements. For compensation to be effective, a reliable assessment is needed of the nature, value and extent of the resource that would be lost, so that like can be replaced with like, or where this is not possible, other related environmental enhancement of at least equal value is undertaken. However, it is arguable whether true compensation is ever possible. For example, a new area of woodland can be created to replace woodland lost to development. In time, this may compensate for the loss of the existing woodland in purely visual terms, but it may never compensate for the loss of established habitat or amenity value. In general, compensation should be regarded as a last resort, and where attempted, expert advice should always be sought, especially on habitat creation.

Available measures

3.74   In areas of high environmental sensitivity, development with unacceptable environmental consequences may not be permitted on these grounds alone. Therefore, it is important that skill and initiative for such locations is employed to find an acceptable form of development by systematic impact mitigation. The most common measures include:

- **Location and siting.** These measures offer the greatest opportunities for effective mitigation, and should be among the first measures considered if landscape, visual and other environmental constraints associated with a particular site prove to be significant.

- **Siting and layout.** Through carefull attention to site layout, even significant environmental constraints can be successfully-accommodated . For example, areas of existing woodland may be incorporated into the essential public open space provision for a major new housing development.

- **Choice of site level.** Landscape is three-dimensional; as well as care with siting and layout, there is also scope for mitigation in the careful choice of site level or vertical alignment.

- Appropriate **form, materials and design of built structures.** Many buildings and structures cannot be screened; nor is it always desirable or practicable. In these circumstances, the design of the structures themselves and their colour treatment and textual finishes, should be considered in such a way that they fit comfortably with their surroundings.

In this example, appropriate colour treatment of the buildings has been used to help the structures fit in with the landscape

- **Ground modelling.** Ground modelling may be undertaken where the natural landform or site levels do not give optimum screening effect. It should be borne in mind that major earthworks in themselves may create adverse landscape and visual impact, and care should be taken to ensure that new landform looks natural and appears as an integral part of the landscape.

- **Planting.** In particular, structural planting, can help to integrate a development with its surroundings, and can soften harsh buildings and structures. Where possible, it should be appropriate to the landscape reflecting local species and be of national provenance. Advance planting and where appropriate, off-site planting, offer special potential for effective mitigation.

3.75 The importance of landscape enhancement, rather than simply mitigation, has yet to be fully recognised. With the twin objectives of economic success and environmental quality now enshrined in government policy on sustainable development, there are signs that an increasingly creative approach to development will be required. The process of mitigating landscape and visual impact is essentially a creative one, and must become more so in future. The aim should be to effect an overall benefit to landscape character and quality, that is a positive impact which can be set against other, negative, environmental impacts. This can be achieved more often than is commonly recognised, particularly where new development can be accommodated in low-grade, despoiled or denuded areas with great potential for improvement.

Presentation of findings

Communication

3.76 The presentation of findings from a landscape and visual impact assessment is important. It should assist decisions on the acceptability of development in landscape and visual terms. A clear structure, plain language and good illustrative and summary materials are vital. The basic aim is to compare the existing landscape with the landscape that will be created before, during and after development.

3.77 **Text** should be concise and to the point. It should be understandable to a lay person, with definitions being given of any technical terms that are used. Criteria and thresholds for sensitivity, magnitude and significance should be carefully stated. The assessment should be written in an impartial style and should focus upon:

- clear description of basic design elements;
- understanding of landscape constraints and opportunities;
- systematic identification of potential impacts;
- sound predictions of impact magnitude;
- reasoned judgements on impact significance;
- measures to address adverse impacts.

3.78 **Illustrations** should be used where they can communicate information more quickly and easily than text. They have a more important role in relation to landscape and visual impacts than any other type of impact: there are many messages that are best communicated through maps, plans, photographs and other illustrative media. The choice of scale and presentational techniques is crucial. Care should be taken to include only information that is relevant to the assessment. For instance, the inclusion of detailed design drawings, such as engineering and architectural technical drawings, may not be appropriate. Illustrations should be closely linked to the text of the assessment, but should complement rather than duplicate its content. It is important to illustrate how the development will relate to both human scale and the scale of the surrounding landscape.

3.79 **Photographs** have a special role in describing landscape character and key views, but can easily be misleading. This may occur if they are taken from unrepresentative viewpoints, or if the framing is such that certain elements are included or excluded. The reasons behind the choice of viewpoints should be given, and the location and direction of view should always be shown on an accompanying map. Seasonal and atmospheric effects and lens type and focal length should also be given. Wide angle lenses reduce the apparent height of landform, buildings and vegetation; while telephoto lenses tend to increase it. Accepted practice is to use a lens with a focal length of between 50 and 65mm; there are inherent dangers in using other lenses by misrepresenting impacts through reduction or enlargement.

3.80 **Charts and tables** tend to be under-used in landscape and visual impact assessment. They can be effective, especially for summarising data on landscape and visual sensitivity, impact magnitude and impact significance. They permit ready comparison between different scheme options and types of impact, and this can be valuable, especially in the early stages of planning and design. In addition, they are probably the best way of making complex information more accessible to consultees and the public. They must be prepared carefully and consistently, as they may often be subject to close scrutiny by decision-makers.

Presentation

3.81 The precise content of a landscape and visual impact assessment may vary considerably, depending on factors such as whether it forms part of a formal or informal EA; the scope of work agreed with the determining authority and consultees; and the sensitivity of the affected landscape and visual resources.

3.82 The general, opening sections of the landscape and visual impact assessment therefore need to present basic information about objectives, responsibilities, methodology and scope. This should include:

- the planning and legal context, for instance, published policies and guidance on landscape designations and landscape character areas in the vicinity of the development;

- the remit, qualifications and experience of those responsible for preparing the assessment;

- the methodology used, including the overall assessment process, the link to scheme design, and the specific techniques used at each stage in assessment;

- the scope of the assessment, key issues, how these were determined and any constraints or data deficiencies that may apply.

3.83   Subsequent sections of the assessment should deal in turn with description of the development; baseline studies; impact identification, prediction and assessment; and mitigation measures. The content of these sections was described in some detail earlier, and hence, the paragraphs that follow concentrate principally on illustrative, photographic and tabular presentation. They suggest the essential presentational materials that might accompany an assessment for a major development scheme. There are a number of presentation techniques that have been developed specifically to help predict and illustrate landscape and visual impacts. These techniques are reviewed at the end of this section.

3.84   **Description of the development.** This section of the assessment, which draws together data about the development proposals, should contain simple, easy-to-read proposals maps at A3 or A4 size, together with other selected drawings. For long and complex projects, such as power stations or major mineral workings, a series of drawings at different stages such as construction, operation, and decommissioning, may be needed. Essential drawings will include:

- plans of the main design elements, including access, layout, land take for different activities, contours and site levels;
- sections and elevations of buildings and other important structures, including key dimensions;
- the proposed basic landscape framework.

3.85   Baseline studies. Baseline studies may need to be presented at more than one scale. Initially they will focus upon the landscape context for the development, especially the enhancement potential and sensitivity of local landscape and visual resources. At this broad scale, presentational materials should include:

- a map of landscape character areas within the zone of visual influence of the development;
- photographs showing the typical appearance of the landscape within each area, and key views;
- a map or diagram showing key issues and priority areas for landscape conservation, enhancement and reconstruction.

3.86   At a more detailed level, there will be a need for:

- maps to show the specific landscape and visual receptors that

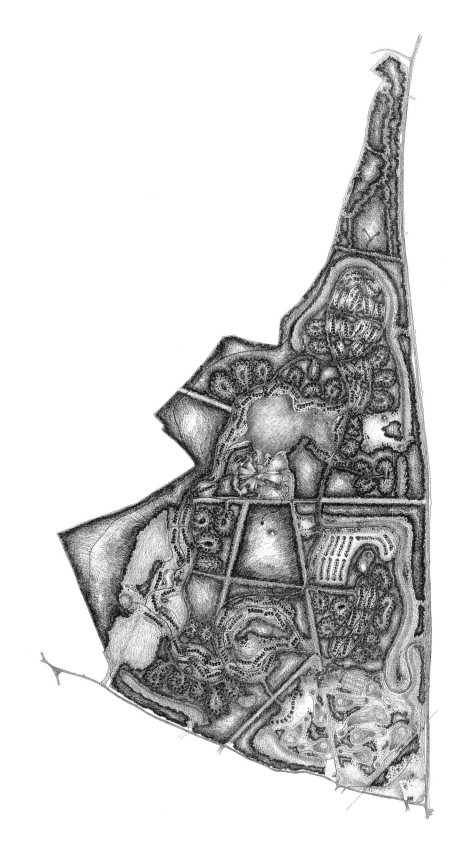

A hand-rendered drawing illustrating the main elements of the proposal

have influenced scheme design.

For the purposes of an ES, such constraints maps may need to be selective, showing only the main landscape elements, features, conservation interests and viewers. They will probably be based on a fuller constraints mapping exercise undertaken for the purposes of scheme design. Different notations may be used to indicate varying degrees of sensitivity.

3.87 **Impact assessment.** Illustrations, photographs and tables can make a particularly useful contribution to this section of the assessment. A very wide range of illustrative techniques is available. Broadly, these can be categorised as:

*   visibility mapping techniques such as zones of visual influence, visual envelopes, visual corridors;
*   visualisations, computer simulations, photomontages, overlays, manually drawn perspectives, sketches, artists' impressions, models and photographs of similar developments.

The former show the extent of landscape and visual impact, and the latter the nature of the impacts that will occur; both are normally essential components of landscape and visual impact assessments. The basic principles, strengths and weaknesses of these key techniques are reviewed below.

3.88 In addition, tabular presentation is very helpful to impact identification, prediction and assessment. Useful tables include:

*   summaries of impact significance for different project options, scheme components or route sections, and stages in the project life-cycle.

3.89 **Mitigation.** Mitigation proposals are generally presented in the form of:

*   a landscape strategy drawing to show how potentially significant landscape and visual impacts have been taken into account in the layout and siting of the development;
*   a landscape masterplan to show how the landscape design will develop effective mitigation measures;
*   sections, in particular to illustrate the reasons for ground modelling or screen planting. Care must be taken over the choice of vertical and horizontal scales so as not to distort the impression conveyed.

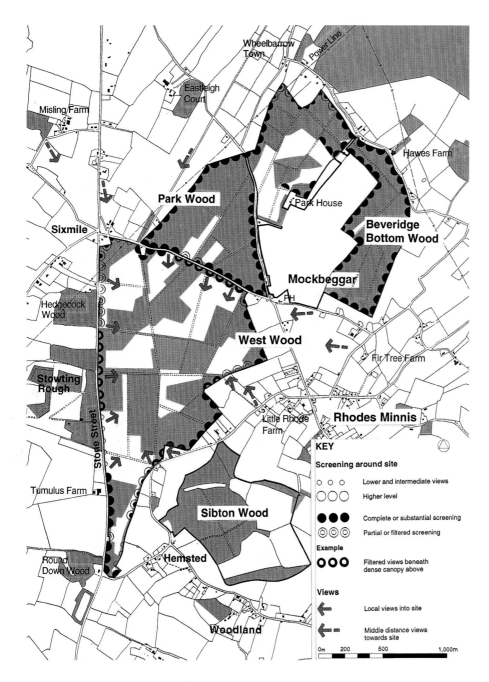

Existing Screening Around Site

A method of using visibility mapping techniques to illustrate the grading of views into the site. In this instance these are taken at the proposed site boundary

A hand rendered drawing showing the project proposals within the context of the surrounding landscape

3.90   The same presentational techniques that were used for baseline studies and impact prediction should be used to illustrate mitigation. The situation both before and after mitigation should be shown, illustrating the effectiveness of mitigation and showing the residual impacts that will prevail after mitigation.

3.91   Where major planting is envisaged, changes in the effectiveness of mitigation through time should be shown, usually through a series of photomontages of the same view before development, at day of opening, at year 5, 10, and at year 15. Assumed growth rates for new planting should be specified.

Presentation techniques

3.92   Visibility mapping and visualisation techniques are central to the effective prediction and communication of landscape and visual impacts. Techniques must be carefully chosen and rigorously applied, as they will be subject to close scrutiny and, in the case of contentious developments, may need to be explained and substantiated at a public inquiry.

3.93   Visibility mapping. In preparing a visibility map, the first issue to consider is whether one seeks to show the visibility of the site itself, or of the development within it. Both are often relevant as visibility of the site will often contribute to the visual amenity of the area. In addition, for complex developments, there may be a need to show separately the areas from which:

*   the whole development will be visible;
*   only part of the development will be visible, for instance, the top of a tall structure, or some of the turbines from a wind farm.

3.94   Manual estimation of visibility from topographic maps is possible, but for developments of any size or complexity, it is strongly recommended that visibility be mapped by computer. A number of specialist computer aided design (CAD) software packages now offer this facility. However, when using a ZVI to measure the extent of visibility their use can be limited, for instance, in a flat landscape where visibility is determined by land use features. For distant views, in theory, allowances must be made for the curvature of the earth's surface and refraction effects of the earth's atmosphere, however, for all but the tallest structures, this would not be a practical consideration.

Illustration showing hand rendering techniques imposed on photographs showing the development of screen planting over a chosen timescale. In this instance 1 and 10 years were selected

3.95   It is essential for visibility to be checked and confirmed in the field, because of the localised screening effects of built elements, minor landform features and vegetation. Summer and winter effects may need to be examined separately to ensure that the worst case is known. Accurate estimation in the field of the visibility of tall structures poses special problems. There are ways of dealing with this problem, such as flying a balloon and using scaffolds and cranes to act as a reference point at the same height and location as the proposed structure, but the practicality of such measures must obviously be taken into account.

3.96   **Visualisations.** Visualisations are one of the best means of communicating the landscape and visual impacts of a development to decision-makers and the public, but again, accuracy is essential. Viewpoints should be chosen with the utmost care. The location, season and timing relative to the project life-cycle should represent conditions of maximum adverse impact.

3.97   A growing range of visualisation techniques is available. The cost, time and skill requirements of each vary markedly. At the top end of the range are computer simulations, built up from digital terrain mapping and air photograph data. These have the ability to present colour images of the development and its surroundings. Once input, data can be manipulated to show different distances and angles of view, or even to give an impression of movement through the landscape. These techniques have great potential, especially in relation to linear developments such as roads and transmission lines. Once baseline conditions are modelled, variations to a scheme can be relatively easily compared, but the cost of the hardware and data input at present is high.

3.98   Photomontages are a popular visualisation technique. Their main advantage is that they show the development within the real landscape and from known viewpoints. Technically, good, accurate photomontages are quite difficult to prepare, and may require specialist advice. First, the development must be placed correctly within the photograph. This requires information on the precise locations and dimensions of the development and known reference points; the accurate location and height from which the photograph was taken; as well as the focal length and direction of view of the camera. Second, for built developments, a computer-generated perspective of key structures will probably be required. Various CAD systems can be used here. Lastly, considerable graphic skill is needed to alter photographs to show realistically new buildings, ground form and planting.

An example of a
computer generated
photomontage

An example of a hand
painted photomontage

3.99   Other visualisation techniques are generally less quantitative
and credible, but may be appropriate under certain circumstances.
The possibilities include overlays, perspective sketches, which are
often constructed over computer generated wire lines; physical models, which tend to be expensive, but are very useful in public consultations, and photographs of similar developments, which are cheap
and can be remarkably helpful, provided it is made clear that they
are indicative only. Artists impressions which are not accurately constructed should be avoided.

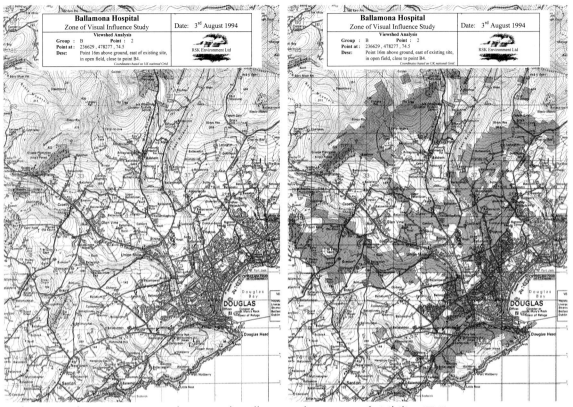

In this example, an acetate overlay is used to illustrate the extent of visibility (ZVI)

Part Four

Consultation, review and implementation

Consultation and the assessment process

4.1   Appropriate consultation is usually an *essential* part of the landscape and visual impact assessment process, without which the effectiveness of the process as a whole may be diminished. Consultation may take a variety of forms and fulfil many purposes at different stages in the assessment. It offers the opportunity to gain advice and help from a wide range of organisations, individuals, local communities and interest groups on a formal and informal basis. It may involve the determining authority, statutory consultees, amenity and conservation bodies and local residents.

4.2   In the initial stages of project planning, consultation with the planning or other determining authority aims to establish the likely acceptability of the proposed development and the preferred development site. It should also identify the need for an assessment through application and interpretation of the appropriate EA Regulations (screening). When an EA is required or when an applicant voluntarily decides to prepare an ES and advises the determining authority of this decision, the determining authority will inform the statutory consultees. This places them under an obligation to provide, on request, information which is likely to be relevant to the assessment.

4.3   Statutory consultees that may hold useful information for landscape and visual impact assessments include the county and district planning authorities, the Countryside Commission, Countryside Council for Wales, Scottish Natural Heritage, English Nature, English Heritage, the National Rivers Authority and the Ministry of Agriculture, Fisheries and Food, (or agencies in Wales, Scotland and Northern Ireland). As well as providing information, for instance, on previous landscape surveys, assessments of the affected area, and advice on the basis for landscape and other designations, consultations with these authorities should aim to arrive at consensus on matters such as terms of reference, methodology and assessment techniques.

4.4   Informal consultations with national and local amenity and conservation groups that have an interest in the project site, can assist the scoping stage of the assessment by giving an early indication of key issues that need to be addressed, and by highlighting useful sources of information and expertise. Such groups may include the local branch of national conservation organisations, eg Council for the Protection of Rural England and the Royal Society for the Protection of Birds, the county wildlife trust, local historical societies and civic organisations. Local schools, parish councils and community groups can also assist, for instance by drawing attention to popular local walks and beauty spots.

4.5   Later in the assessment process, consultation will be more concerned with information dissemination, particularly to the determining authority and statutory consultees. The interim presentation of information about the project to interested parties helps to maintain goodwill. The timing of such involvement will depend upon the nature of the development, but generally, the earlier this occurs, the better. If possible, a two-way exchange of ideas and views on scheme design should be initiated. Such participation may not always result in full agreement, but at the very least it should serve to resolve some issues and to clarify any objections which remain. In its most useful form, participation will improve the quality of the information that influences scheme design, and should result in positive changes to that design.

4.6   The preparation of a **Non-Technical Summary** (NTS) is required by the EA legislation and is expressly intended to help the general public and interested parties to participate in the decision making process from an informed position. The NTS should be well illustrated with photographs and easily understood plans. Summary descriptions should be cross-referenced back to the full ES so that the general public can refer to particular areas for more detail if required.

Guidelines for consultation

4.7   Consultation is not without its difficulties. Often the developer is reluctant to release information about the development which may be commercially sensitive. There is also a perception that to invite discussion and debate is to open the project to interference. The adverse publicity generated by a poorly handled programme of consultations may be considerable.

4.8   From the point of view of statutory consultees and members of the public, there may be suspicion about the consultation process. They may not be convinced that the offer to participate is genuine, and may be afraid that they will prejudice their future position if they become too closely involved.

4.9   Most of these reservations can be overcome if those who are handling the consultation programme are guided by a few simple rules:

- **Consultation should be genuine and open.** The temptation to make the most of consultation for information gathering while being reluctant with information dissemination should be resisted.

- **The timing of consultation should be carefully planned** to prevent premature disclosure which may encourage blight or make developers commercially vulnerable. Although potential developers should not be oversensitive on this matter, there may be occasions where controlled release and confidentiality safeguards are required.

- **Requests for participation should be timely.** There is no point in seeking ideas and views if it is actually too late for scheme design to be modified. Sufficient time must also be allowed for consultees to be able to consider and act on the information provided.

- **Developers should be clear as to the issues on which comment is, and is not, being sought.** The objectives of consultation should be clearly stated. Information presented to consultees should be appropriate in content and level of detail.

Consultation methods

4.10   The objectives and the stage in the assessment process will dictate the method or combination of methods used for consultation. There are both formal and informal consultation elements with the UK EA system, but informal and early consultations hold by far the greatest potential. Statutory consultations, which come after the application has been submitted, almost always occur too late to have a real influence upon scheme design or prevention of landscape and visual impact. The main consultation methods that can be used are described below.

4.11   **Face-to-face discussion.** Much of the informal consultation will be through direct discussion with relevant parties, particularly the planning authority, conservation agencies and other appropriate statutory consultees. For major developments it may be helpful to establish an environmental liaison or advisory group. This can make the consultation process less time-consuming and more cost-effective, as it should assist the various organisations in arriving at a consensus view. Early joint site visits may also be very helpful as a means of exploring landscape opportunities and constraints.

4.12   **Correspondence.** Correspondence may be used for information gathering and dissemination, to invite comment, or to record issues that have been discussed and agreed. Care should be taken to state the purpose of the correspondence.

4.13   **Presentations and informal public meetings.** There may be occasions when select presentations, for example to elected members, may be useful. In general, though, select presentations should be avoided because of the risk of alienating excluded parties. Open public meetings are occasionally useful to assist public participation. However, views arising from such meetings are often hard to record. Worse still, they can get out of control and result in fruitless confrontation.

4.14   **Exhibitions.** Exhibitions can present a great deal of information to a broad range of parties, and if successful, can offer an opportunity for friendly and constructive exchange of views between the design and assessment team and members of the public who will be directly affected. It is good practice for developers to consult the general public, and this is one of the best ways of doing so.

Public exhibitions can be one of the best ways of presenting large amounts of information on the proposed development to the public

4.15   **Workshops.** Workshops are highly structured public meetings in which those attending are expected to participate and produce results, thereby encouraging involvement in the decision making process. They can be used at many stages in the EA process but may be particularly useful for identifying key issues and concerns and for examining options for impact mitigation. The selection of participants will be crucial and all interests should ideally be represented to prevent feelings of exclusion. However, to be effective the group size should be limited to around 20-25 people, with an optimum size for small groups of 5-7 people.

4.16 **Leaflets and mailings.** Lastly, consultation material may also be presented to the public through leaflets or mailings to representative groups or to all affected local residents. This approach is often used in conjunction with public exhibitions, and may include a written response to a questionnaire. In relation to landscape and visual impacts, views can be sought on preferred landscape enhancement measures and on the form that mitigation should take to meet local needs.

Examples of mailings for public consultation

4.17   Consultation has an important part to play in landscape and visual impact assessment, but requires a commitment from the developer and the consultee to respond positively and constructively. It is a valuable tool for reaching mutual understanding and agreement on the key issues of landscape and visual impact, which are often subjective. It can highlight local interests and values, which are easily overlooked. Most importantly, it can help to generate a vision of the ideal landscape after development, and hence make a real contribution to scheme design.

The role of the determining authority

How the authority can help

4.18   This section is directed at the determining authority, and to a lesser extent at those statutory consultees who have a strong interest in landscape issues. It contains advice on:

* how to encourage the best result from the developer;
* how to review the landscape and visual component of an ES and judge whether or not it is adequate;
* implementation and monitoring of scheme proposals.

4.19   It will be in the interests of the general public, as well as the developer, if this role can be undertaken with the appropriate professional expertise. Otherwise it will be difficult to ensure that the determining authority is able to scope adequately, to accredit work, or to generate the necessary rise in demands for high quality environmental considerations during the whole development process.

Guidance to the developer

4.20   There are a number of different ways in which planning authorities and others can give support and guidance to those responsible for the planning, design and assessment of new developments and at the same time further their own policies, interests and objectives.

4.21   **Landscape context and design principles.** The general planning context for new development is contained in formal, adopted structure plan and local plan policies which indicate in broad terms what types of development will be acceptable where. However, it can be enormously helpful to the development process if planning authorities prepare additional, informal advice on landscape matters. This is encouraged by the DoE in PPG 7 *The Countryside and the Rural Economy* [13].

4.22   Area-wide landscape assessments and strategies can provide a clear framework for the consideration of landscape issues in development control and allow local authorities to be proactive and guide development to the most suitable areas. They can advise on existing landscape character; on areas identified for landscape conservation, restoration or reconstruction; and on key issues that need to be addressed, such as urban fringe landscape degradation and industrial dereliction. At a more detailed level, landscape design guidance can help encourage local distinctiveness in building layout, style, materials, and choice of plant species. For example, the Countryside Commission has recently issued advice on both landscape assessment [7] and design in the countryside [14] and this gives further ideas on the type of guidance that may be appropriate.

4.23 **Content of the landscape and visual impact assessment.** It is good practice for the determining authority to work closely with the developer on the terms of reference, scope and methodology for the landscape and visual impact assessment. The authority should also provide guidance on the level of detail that would be appropriate to the assessment. Some authorities have begun to issue standard advice on these matters, including:

- checklists of information held by the authority, and additional baseline survey information that will be required;
- advice on environmental constraints, for instance, on landscape designations;
- general principles and objectives regarding landscape enhancement and mitigation measures.

4.24 This advice can be helpful to developers, especially at an early stage in the assessment process. However, it is no substitute for discussions and advice on specific requirements for a given development proposal and site. For instance, it may be helpful for the authority to review and agree baseline survey findings; criteria and thresholds for sensitivity, magnitude and significance; and 'performance standards' for mitigation of key impacts.

4.25 **Design proposals.** The developer should be given guidance on the level of detail of the landscape design proposals that should be submitted in support of the development application. Guidance may cover:

- protection of existing features such as trees, shrubs, watercourses, areas of nature conservation interest, historic landscape features;
- content of landscape proposals for instance scale, level of detail, programme for implementation;
- planting design, including off-site and advance planting;
- earthworks, mounding and contouring;

- surface water drainage;
- floodlighting;
- overground and underground services;
- fencing and boundary treatments.

4.26   In addition, advice should usefully cover management, maintenance, restoration and other matters on which the developer should make a clear long term commitment. This requires the preparation of maintenance schedules, landscape management plans and in the case of minerals and waste disposal sites, phased programmes of working and restoration. Model and specific conditions should be drafted and discussed.

Review of the landscape and visual impact assessment

4.27   The second main function of the determining authority is to review the adequacy of the landscape and visual impact assessment, including collation and consideration of relevant comments from statutory consultees. The review process involves checking that the assessment meets the requirements of the EA Regulations and also the specific terms of reference discussed and agreed with the developer. Before undertaking the review, the authority should satisfy itself that it is competent to carry out the work in-house. In the absence of in-house landscape staff, it may be necessary to seek specialist advice from outside.

4.28   The review should check: the scope and content of baseline studies; methodology and techniques applied; accuracy and completeness of impact identification and prediction; criteria and thresholds used to assess impact significance; effectiveness of proposed mitigation; and success in communicating results. The IEA has developed a set of general criteria for reviewing ESs [6] that can also be applied to the landscape and visual impact assessment section of the ES. In addition, the University of Manchester, in work published by the Countryside Commission [8] and endorsed by the Countryside Council for Wales, has developed specific criteria for review of landscape impact assessments. Key checks that should be carried out when reviewing a landscape and visual impact assessment are given in the box below.

4.29   If the landscape and visual impact assessment is found to provide insufficient information for proper consideration of the application, the determining authority may ask the developer to supply additional information.

Is the baseline information adequate and comprehensive? For example are all topics covered?

Is the existing dynamic of the landscape adequately described?

Is the baseline information correct?
e.g.   Is mapping data up-to-date?

Is the sensitivity to change of landscape and visual resource clearly defined, for example, by reference to recognised national, regional or local importance, professional judgement or scenic quality, special conservation interests and known public preferences?

Does it take into account seasonal/climatic variations?
e.g.   What water table fluctuations have been monitored?

Does it take into account the appropriate codes, standards and best practice?
e.g.   Have latest standards been used?

Description of the Development

Is the description of the project adequate?
e.g.   Is the project described in language which is easy to understand?

Is the development described so it can be viewed in its wider landscape context?

Is the description of the project correct in all respects?
e.g   Has anything been omitted?

Does it include sufficient detail to determine the proposal during:
a)   Construction
b)   Operation
c)   Future Maintenance/Management?

Is sufficient attention given to off-site issues?
e.g.   Off-site planting

Identification, Prediction and Assessment

Qualitative and quantitative predictions of impact magnitude should be made and best practice should be adopted for assessment of impact significance.

What is the methodology used and is it appropriate?

Are the assessment methods and techniques adequate?

Are these appropriate to the particular kind of assessment?

Are all the impacts identified and does the assessment provide a full and objective account of all direct and indirect impacts:
a)   Temporarily

b)   Short Term
c)   Medium Term
d)   Long Term
e)   Cumulative?

Does the assessment consider adequately both positive and negative aspects of the proposal?
e.g.   How effectively and accurately does it predict impacts?

Does the assessment consider fully and without bias all alternatives and have these been adequately evaluated?

Mitigation

Proposed mitigation measures should address all key landscape and visual impacts adequately and effectively.

Does the assessment provide an unbiased opinion of adverse effects?
e.g.   Are the conclusions realistic?

Are the measures proposed for this mitigation adequate and what is the impact of these measures?

Are they likely to be:

a)   feasible e.g. formal agreements with adjacent landowners
b)   effective e.g. how long will it take to implement and what provision is made for
     management/maintenance
c)   acceptable e.g. within the framework of the existing
     landscape?

Has sufficient consideration been given to timescale?
eg.   Short term impacts of materials such as geosynthetics.

What alternative measures have been considered and why have these been dismissed?

Presentation of Findings

Findings should be balanced, clear, concise and well-illustrated with a minimum of technical jargon. For example:

Is the information presented clearly?

Is the graphic material of good quality, accurate and relevant?

Is statistical information easy to understand?

Do the main sections of the report demonstrate adequate interaction and cross-referencing between different subjects?

4.30   The granting of development consent is not an end in itself. The developer, advisers and the determining authority all have a responsibility to ensure that commitments made in the application and the ES are honoured during construction, operation, subsequent site management and restoration.   Usually implementation is achieved through the enforcement of consent conditions, legal agreements, undertakings, or other requirements.

4.31   To be effective, these must be relevant, fair, reasonable and enforceable. These can be based in part on standard conditions such as those outlined above, but should be tailored to reflect the partic-ular impacts of the scheme concerned. For instance, in relation to highways development, the most important undertakings will prob-ably relate to ground modelling and screen planting; while planning conditions for mineral extraction proposals will focus on phased working, progressive restoration, and aftercare of restored areas. Where the effectiveness of proposed mitigation measures is uncer-tain, performance standards may be applied describing what should be achieved rather than how, and appropriate checking procedures should be built in. In certain cases a financial bond may be appro-priate, which would be held until mitigation measures have been suc-cessfully completed.

4.32   Mitigation measures require long term management and monitoring. In addition monitoring can fulfil a number of very use-ful purposes. It can:

- establish whether or not predicted impacts have actually occurred;
- identify unforeseen impacts and omissions from the original ES;
- check compliance with proposed mitigation measures and plan-ning conditions;
- check the effectiveness of mitigation measures in avoiding or reducing adverse impacts.

4.33   There are several advantages to be gained from monitoring impacts in this way. With appropriate feedback to similar future assessments, the quality of impact prediction can be improved through time. Remedial action can sometimes be taken to address unforeseen impacts. Lastly, enforcement action can be carried out where necessary to ensure that the mitigation measures that were promised are implemented, and are effective. For all these reasons, the development of explicit monitoring programmes for landscape and visual impacts is strongly encouraged.

4.34   The responsibility for monitoring must lie jointly with the developer and the determining authority. For the developer, as well as ensuring the successful outcome of the project, monitoring can enhance credibility and public confidence. For the planning or other authority, monitoring offers the opportunity to check on the effectiveness of mitigation and take appropriate action to ensure that landscape conservation and enhancement objectives have in fact been achieved. Finally, monitoring can help to improve the future practice of landscape and visual impact assessment, by providing feedback on the accuracy of assessment techniques.

Abbreviations

AGLV	Area of Great Landscape Value
AONB	Area of Outstanding Natural Beauty
CAD	Computer Aided Design
DoE	Department of Environment
EA	Environmental Assessment
EC	European Commission
ES	Environmental Statement
IEA	Institute of Environmental Assessment
NTS	Non-Technical Summary
PPG	Planning Policy Guidance
NSA	National Scenic Area
VEM	Visual Envelope Map
ZVI	Zone of Visual Influence

Glossary

Analysis, (landscape)	The process of breaking the landscape into its component parts to understand how it is made up.
Assessment	An umbrella term for description, analysis and evaluation.
Landscape Character	A distinct pattern or combination of elements that occurs consistently in parts of the landscape.
Character area	A geographic area with a distinctive character.
Determining Authority	The planning or other authority responsible for planning consents or project authorisation.
Constraints map	Map showing the location of sensitive receptors.
Do nothing situation	Continued change/evolution of landscape or of the environment in the absence of the proposed development.
Element	A component part of the landscape, eg. roads, hedges, and woods.
Enhancement	Landscape improvement through restoration, reconstruction or creation.
Environmental fit	The relationship of a development to identified environmental opportunities and constraints.
Landscape feature	A prominent eye-catching element, eg. wooded hill top, and church spire.
Indirect impact	Impacts which occur as a secondary or tertiary effect of a development.
Landform	Combination of slope and elevation producing the shape and form of the land surface.

Landscape evaluation	The process of attaching value (non-monetary) to a particular landscape, usually by reference to an agreed set of criteria and in the context of the assessment.
Landscape impacts	Change in the fabric, character and quality of the landscape as a result of development. These can be positive or negative.
Magnitude	Size, extent and duration of an impact.
Method	The specific approach and techniques used for a given study
Mitigation	Measures designed to avoid, reduce, remedy or compensate for landscape and visual impacts.
Precautionary Principle	Principle applied to err on the side of caution when the prediction of environmental impacts is uncertain.
Landscape Quality	Term used to indicate value based on character, condition and aesthetic appeal.
Landscape Resource	The combination of elements that contributes to landscape context, character and value.
Scoping	The process of identifying the potentially significant impacts of a development.
Sense of place (or Genius Loci)	The essential character and spirit of an area. Genius Loci literally `spirit of the place'.
Sensitive Receptor	Physical or natural landscape resource, special interest or viewer group that will experience an impact.
Sensitivity	Vulnerability of sensitive receptor to change.
Technique	Specific working process.
Impact Threshold	A specified level of impact significance.

Tolerance — The ability of a landscape or view to accommodate change.

Visual amenity — The value of a particular area or view in terms of what is seen.

Visual impact — Change in the appearance of the landscape as a result of development. This can be positive (improvement) or negative (detraction).

Visual intrusion — Degree to which a development intrudes upon the field of view.

Visualisation — Computer simulation, photomontage or other technique to illustrate the appearance of a development.

Worst case situation — Principle applied where the environmental impacts may vary eg. seasonally, to ensure the most severe impact is assessed.

Zone of visual influence/ visual envelope — Extent of visibility to or from the development.

1   Council of the European Communities (1985), Council Directive of 27 June 1985 on the assessment of the effects of certainpublic and private projects on the environment (85/337/EEC), *Official Journal of the European Communities*, 5.7.85, L175/40-48.

2   Department of the Environment and Welsh Office (1989), *Environmental Assessment: A Guide to the Procedures*, HMSO, London.

3   Department of the Environment (1990), *This Common Inheritance*, HMSO, London.

4   Department of the Environment (1994), *Sustainable Development: The UK Strategy*, HMSO, London.

5   Wood, C and Jones, C(1991), *Monitoring Environmental Assessment and Planning*, Department of the Environment Planning Research Programme, HMSO, London.

6   Coles, TF and Tarling, JP (1991), *Environmental Assessment: Experience to Date*, Institute of Environmental Assessment, Lincolnshire.

7   Countryside Commisssion (1993), *Landscape Assessment Guidance*, CCP3 423, Countryside Commission, Cheltenham.

8   Countryside Commission (1991), *Environmental Assessment: The Treatment of Landscape amd Countryside Recreation Issues*, CCP 326, Countryside Commission, Cheltenham.

9   Countryside Commission, English Heritage, English Nature (1993), *Conservation Issues in Strategic Plans*, CCP 420, Countryside Commission, Cheltenham.

10   Department of the Environment (1993), *Environmental Appraisal of Development Plans: A Good Practice Guide*, HMSO, London.

11   Land Use Consultants (1991), *Landscape Assessment: Principles and Practice,* Coutryside Commission for Scotland, Battleby, Perth.

12   Countryside Council for Wales (1993), *Interim Guidance on Landscape Assessment,* Countryside Council for Wales, Bangor.

13   Department of the Environment and Welsh Office (1992), *The Countryside and the Rural Economy,* PPG 7, HMSO, London.

14   Countryside Commission (1993), *Design in the Countryside,* CCP 418, Countryside Commission, Cheltenham.

15   World Commission on Environment and Development (1987), *Our Common Future,* Oxford University Press, Oxford.

Bourassa, S C (1991), *The Aesthetics of Landscape*, Belhaven Press, London.

Cheshire County Council (1989), *The Cheshire Environmental Assessment Handbook*, PPN No.2, Cheshire County Council.

Countryside Commission (1991), *Wind Energy Development and the Landscape*, CCP 357, Countryside Commission, Cheltenham.

Countryside Commission (1988), *A review of recent practice and research in landscape assessment*, Countryside Commission, Cheltenham.

Department of Transport, Scottish Office Industry Department, The Welsh Office, the Department of the Environment for Northern Ireland (1993), *Design Manual for Roads and Bridges, Volume 11: Environmental Assessment*, HMSO, London.

Heape, M (1991), *Visial Impact Assessment*, unpublished paper presented to the 12th International Seminar on Environmental Assessment, University of Aberdeen.

Meetham, R (1993), *Landscape and Visual Impact Assessment*, unpublished MSc thesis, University of Wales at Aberystwyth.

National Rivers Authority (1993), *River Landscape Assessment, Conservation Technical Handbook 2*, National Rivers Authority, Bristol.

Price, G (1993), *Landscape Assessment for Indicative Forestry Strategies*, Forestry Authority England, Cambridge.

Sheppard, S R J (1989), *Visual Simulation*, Van Nostrand Reinhold, New York.

U S Department of the Interior (1977), *BLM Manual, Sections 6300, 6310 and 6320, Visual Resource Management*, Bureau of Land Management, Washington DC.

Appendix One

The EA process

The EA process

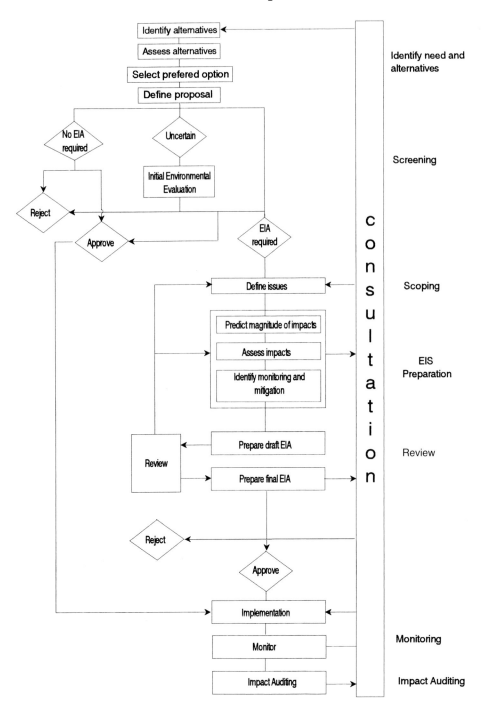

Flow diagram of the main components of the EA process
(Based on Wathern 1988)

Appendix Two

Projects requiring environmental assessment (summary of Schedule 1 and 2 projects and the DoE's indicative criteria and thresholds for Schedule 2 projects)

Schedule 1 projects

The following types of development ("Schedule 1 projects") require environmental assessment in every case:

(1)   The carrying out of building or other operations, or the change of use of buildings or other land (where a material change) to provide any of the following:

1   A crude-oil refinery (excluding an undertaking manufacturing only lubricants from crude oil) or an installation for the gasification and liquefaction of 500 tonnes or more of coal or bituminous shale per day.

2   A thermal power station or other combustion installation with a heat output of 300 megawatts or more, other than a nuclear power station or other nuclear reactor.

3   An installation designed solely for the permanent storage or final disposal of radioactive waste.

4   An integrated works for the initial melting of cast-iron and steel.

5   An installation for the extraction of asbestos or for the processing and transformation of asbestos or products containing asbestos:

(a)   where the installation produces asbestos-cement products, with an annual production of more than 20 000 tonnes of finished products; or

(b)   where the installation produces friction material, with an annual production of more than 50 tonnes of finished products; or

(c)   in other cases, where the installation will utilise more than 200 tonnes of asbestos per year.

6   An integrated chemical installation, that is to say, an industrial installation or group of installations where two or more linked chemical or physical processes are employed for the manufacture of olefins from petroleum products, or of sulphuric acid, nitric acid, hydrofluoric acid, chlorine or fluorine.

7   A special road; a line for long-distance railway traffic; or an aerodrome with a basic runway length of 2100m or more.

8   A trading port, an inland waterway which permits the passage of vessels of over 1350 tonnes or a port for inland waterway traffic capable of handling such vessels.

9   A waste-disposal installation for the incineration or chemical treatment of special waste.

(2)   The carrying out of operations whereby land is filled with special waste, or the change of use of land (where a material change) to use for the deposit of such waste.

Schedule 2 projects

The following types of development ("Schedule 2 projects") require environmental assessment if they are likely to have significant effects on the environment by virtue of factors such as their nature, size or location:

1   Agriculture

(a)   water management for agriculture
(b)   poultry-rearing
(c)   pig-rearing
(d)   a salmon hatchery
(e)   an installation for the rearing of salmon
(f)   the reclamation of land from the sea

2   Extractive industry

(a)   extracting peat
(b)   deep drilling, including in particular:
    (i)   geothermal drilling
    (ii)   drilling for the storage of nuclear waste material
    (iii)   drilling for water supplies
    but excluding drilling to investigate the stability of the soil
(c)   extracting minerals (other than metalliferous and energy-producing minerals) such as marble, sand, gravel, shale, salt, phosphates and potash
(d)   extracting coal or lignite by underground or open-cast mining
(e)   extracting petroleum
(f)   extracting natural gas

(g)   extracting ores

(h)   extracting bituminous shale

(i)   extracting minerals (other than metalliferous and energy-producing minerals) by open-cast mining

(j)   a surface industrial installation for the extraction of coal, petroleum, natural gas or ores or bituminous shale

(k)   a coke oven (dry distillation of coal)

(l)   an installation for the manufacture of cement

3   Energy industry

(a)   a non-nuclear thermal power station, not being an installation falling within Schedule 1, or an installation for the production of electricity, steam and hot water

(b)   an industrial installation for carrying gas, steam or hot water; or the transmission of electrical energy by overhead cable

(c)   the surface storage of natural gas

(d)   the underground storage of combustible gases

(e)   the surface storage of fossil fuels

(f)   the industrial briquetting of coal or lignite

(g)   an installation for the production or enrichment of nuclear fuels

(h)   an installation for the reprocessing or irradiated nuclear fuels

(i)   an installation for the collection or processing of radioactive waste, no being an installation falling within Schedule 1

(j)   an installation for hydroelectric energy production

(k)   a wind generator

4   Processing of metals

(a)   an ironworks or steelworks including a foundry, forge, drawIng plant or rolling mill (not being a works falling with in Schedule 1)

(b)   an installation for the production (including smelting, refining, drawing and rolling) of non-ferrous metals, other than precious metals

(c)   the pressing, drawing or stamping of large castings

(d)   the surface treatment and coating of metals

(e)   boilermaking or manufacturing reservoirs, tanks and other sheetmetal containers

(f)   manufacturing or assembling motor vehicles, or manufacturing motor-vehicle engines

(g)   a shipyard

(h)   an installation for the construction or repair of aircraft

(i)   the manufacture of railway equipment

(j)   swaging by explosives

(k)   an installation for the roasting or sintering of metallic ores

5   Glass making

the manufacture of glass

6   Chemical industry

(a)   the treatment of intermediate products and production of chemicals, other than development falling within Schedule 1

(b)   the production of pesticides or pharmaceutical products, paints or varnishes, elastomers or peroxides

(c)   the storage of petroleum or petrochemical or chemical products

7   Food industry

(a)   the manufacture of vegetable or animal oils or fats

(b)   the packing or canning of animal or vegetable products

(c)   the manufacture of dairy products

(d)   brewing or malting

(e)   confectionery or syrup manufacture

(f)   an installation for the slaughter of animals

(g)   an industrial starch manufacturing installation

(h)   a fish-meal or fish-oil factory

(i)   a sugar factory

8   Textile, leather, wood and paper industries

(a)   a wood scouring, degreasing and bleaching factory

(b)   the manufacture of fibre board, particle board or plywood

(c)   the manufacture of pulp, paper or board

(d)   a fibre-dyeing factory

(e)   a cellulose-processing and production installation

(f)   a tannery or a leather dressing factory

9   Rubber industry

the manufacture and treatment of elastomer-based products

10   Infrastructure projects

(a)   an industrial estate development project
(b)   an urban development project
(c)   a ski-lift or cable-car
(d)   the construction of a road, or a harbour, including a fishing harbour, or an aerodrome, not being development falling within Schedule 1
(e)   canalisation or flood-relief works
(f)   a dam or other installation designed to hold water or store it on a long term basis
(g)   a tramway, elevated or underground railway, suspended line or similar line, exclusively or mainly for passenger transport
(h)   an oil or gas pipeline installation
(i)   a long distance aqueduct
(j)   a yacht marina
(k)   a motorway service area
(l)   coast protection works

11.   Other projects

(a)   a holiday village or hotel complex
(b)   a permanent racing or race track for cars or motor cycles
(c)   an installation for the disposal of controlled waste or waste from mines and quarries, not being an installation falling within Schedule 1
(d)   a waste water treatment plant
(e)   a site for depositing sludge
(f)   the storage of scrap iron
(g)   a test bench for engines, turbines or reactors
(h)   the manufacture of artificial mineral fibres
(i)   the manufacture, packing, loading or placing in cartridges of gunpowder, or other explosives
(j)   a knackers' yard
12   The modification of a development which has been carried

out, where that development is within a description mentioned in
Schedule 1.

13   Development within a description mentioned in Schedule 1,
where it is exclusively or mainly for the development and testing of
new methods or products and will not be permitted for longer than
one year.

Indicative criteria and thresholds for identifying Schedule 2 projects requiring Environmental Assessment

It should be borne in mind that the fundamental test to be applied in each case is the likelihood of significant environmental effects. Factors stated in determining significance are given as the nature, size or location of the development.

It is stressed that this is a summary of the guidance and the reader is referred to the appropriate circulars and guidance (see reference 3).

Projects which may require an EA:

Agriculture

New pig-rearing installations - more than 400 sows or 5000 fattening pigs.

New poultry-rearing installations - more than 100 000 broilers or 50 000 layers, turkeys or other poultry.

Salmon farming - production of more than 100 tonnes of fish a year depending on the environmental effects generally and on the river system.

New drainage and flood defence works - may merit EA where it emerges from the consultations between drainage bodies and environmental interests that the project in question is likely to have a significant environmental effect.

Extractive Industry

Depends on such factors as sensitivity of the location, size, working methods, the proposals for disposing of waste, the nature and extent of processing and ancillary operations and arrangements for transporting minerals away from the site. The duration of the proposed workings is also a factor to be taken into account.

Minerals applications in national parks and AONBs should generally be subject to EA.

All new deep mines, apart from small mines, may merit EA.

Opencast coal mines and sand gravel workings more than 50ha or if in a sensitive area may well require an EA.

Whether rock quarries or clay operations or other mineral working require an EA depends on the location, scale and type of activities proposed.

Oil and gas extraction - depends on volume, transportation and sensitivity of area affected. An EA may be required where production is expected to be substantial (more than 300 tonnes per day), or the site concerned is sensitive to disturbance from normal operations.

Energy Industry

Wind generators may require EA - if the development is located within or have potentially significant effects on a National Park, the Broads, the New Forest, an AONB, SSSI or heritage coast; or more than ten wind generators are proposed, or the total capacity is over 5 megawatts.

Manufacturing Industry

Projects may require EA where:

New manufacturing plants are in the range of 20-30ha and above.

New plants may occasionally require EA on account of the expected discharge of waste, emission of pollutants etc.

Industrial Estate Development Projects

An EA may be required where the site area is in excess of 20ha or is in close proximity to significant numbers of dwellings e.g. more than 1000 dwellings within 200m of the site boundaries.

Urban Development Projects

An EA may be required for new projects where:

the site area of the scheme is more than 5ha in an urbanised area; significant numbers of dwellings are in close proximity to the site, e.g. more than 700 dwellings within 200 metres of the site boundaries; or the development would provide a total of 10 000m² of shops, offices or other commercial uses. For out-of-town shopping centres the indicative threshold is 20 000m².

Local Roads

An EA may be required where a road is over 10km in length outside urban areas and 1km if the road passes through a national park or through or within 100m of a SSSI, a national nature reserve or a conservation area.

Within urban areas, any scheme where more than 1500 dwellings lie within 100m of the centre line of the proposed road may require an EA.

Airports

Airports with a runway length of over 2100 metres will require an EA under Schedule 1. Smaller new airports will also generally require EA. Major works such as new runways or passenger terminals at larger airports may also require EA.

Other Infrastructure Projects

Projects requiring sites in excess of 100ha may require an EA.

Motorway service areas

An EA may be required where the proposed location is in a National Park, the Broads, the New Forest, an AONB or SSSI, or is over 5ha outside these locations.

Coast Protection Works

An EA may be required where works are to be located in or potentially have significant effects on a National Park, the Broads, the New Forest, an AONB, SSSI, heritage coast or a marine nature reserve.

Waste Disposal

Installations with a capacity of more than 75 000 tonnes per year may require an EA.

Appendix Three

Types of determining authorities for environmental assessment

TYPES OF DETERMINING AUTHORITIES FOR ENVIRONMENTAL ASSESSMENT

1   Department of the Environment

2   Department of Transport

3   Ministry of Agriculture, Fisheries and Foods (for land drainage and forestry decisions)

4   Scottish Office

5   Welsh Office

6   Northern Ireland Office

7   Department of Trade (pipelines etc)

8   County Councils

9   Metropolitan Borough Councils

10   London Boroughs

11   District Councils

12   Development Corporations

13   National Park Authorities

14   Crown Estates Commissioners

15   Houses of Parliament

Appendix Four

Institute of Environmental Assessment review criteria for environmental statements

A   Excellent, no tasks left incomplete
B   Good, only minor omissions and inadequacies
C   Satisfactory despite omissions and inadequacies
D   Parts well attempted, but must as a whole be considered just unsatisfactory because of omissions and/or inadequacies
E   Poor, significant omissions or inadequacies
F   Very poor, important tasks poorly done or not attempted
N/A   Not applicable. The review topic is not applicable or relevant in the context of this statement

INSTITUTE REVIEW CRITERIA

1.0   Description of the development, the local environment and the baseline conditions

1.1   Description of the development

The purpose and objectives of the development should be explained. The description of the development should include the physical characteristics, scale and design as well as quantities of material needed during construction and operation. The operating experience of the operator and the process, and examples of appropriate existing plant, should also be given.

1.2   Site description

The area of land affected by the development should be clearly shown on a map and the different land uses of this area clearly demarcated. The affected site should be defined broadly enough to include any potential effects occurring away from the construction site (eg dispersal of pollutants, traffic, changes in channel capacity of watercourses as a result of increased surface run off etc).

1.3   Residuals

The types and quantities of waste matter, energy and residual materials and the rate at which these will be produced should be estimated. The methods used to make these estimations should be clearly described, and the proposed methods of treatment for the waste and residual materials should be identified. Waste should be quantified wherever possible.

A description of the environment as it is currently and as it could be expected to develop if the project were not to proceed. Some baseline data can be gathered from existing data sources, but some will need gathering and the methods used to obtain the information should be clearly identified. Baseline data should be gathered in such a way that the importance of the particular area to be affected can be placed into the context of the region or surroundings and that the effect of the proposed changes can be predicted.

2.0   Identification and evaluation of key impacts

2.1   Identification of impacts and method statement

The methodology used to define the project specification should be clearly outlined, in a method statement. This statement should include details of consultation for the prepartation of the scoping report, discussion with expert bodies (eg Planning Authority, HMIP, NRA, JNCC, English Nature, Countryside Commission or Scottish Natural Heritage etc.) and the public, and reference to panels of experts, guidelines, checklists, matrices, previous best practice examples of environmental assessments on similar projects (whichever are appropriate). Consideration should be given to impacts which may be positive or negative, cumulative, short or long term, permanent or temporary, direct or indirect. The logic used to identify the key impacts for investigation and for the rejection of others should be clearly explained. The impacts of the development on human beings, flora and fauna, soil, water, air, climate, landscape, material assets, cultural heritage, or their interaction, should be considered. The method statement should also describe the relationship between the promoters, the planning, engineering and design teams and those responsible for the ES.

2.2   Prediction of impact magnitude

The size of each impact should be determined as the predicted deviation from the baseline conditions, during the construction phase and during normal operating conditions and in the event of an accident when the proposed development involves materials that could be harmful to the environment (including people). The data used to estimate the magnitude of the main impacts should be clearly described and any gaps in the required data identified. The methods used to predict impact magnitude should be described and should be appropriate to the size and importance of the projected disturbance.

Estimates of impacts should be recorded in measurable quantities with ranges and/or confidence limits as appropriate. Qualitative descriptions where necessary should be as fully defined as possible (eg "insignificant means not perceptible from more than 100 m distance").

2.3   Assessment of impact significance

The significance of all those impacts which remain after mitigation should be assessed using the appropriate national and international quality standards where available. Where no such standards exist, the assumptions and value systems used to assess significance should be justified and the existence of opposing or contrary opinions acknowledged.

3.0   Alternatives and mitigation

3.1   Alternatives

Alternative sites should have been considered where these are practicable and available to be developed. The main environmental advantages and disadvantages of these should be discussed in outline, and the reasons for the final choice given. Where available, alternative processes, designs and operating conditions should have been considered at an early stage of project planning and the environmental implications of these outlined.

3.2   Mitigation

All significant adverse impacts should be considered for mitigation and specific mitigation measures put forward where practicable. Mitigation methods considered should include modification of the project, compensation and the provision of alternative facilities as well as pollution control. It should be clear to what extent the mitigation methods will be effective. Where the effectiveness is uncertain or depends on assumptions about operating procedures, climatic conditions etc, data should be introduced to justify the acceptance of these assumptions.

3.3   Commitment to mitigation

Clear details of when and how the mitigation measures will be carried out should be given. When uncertainty over impact magnitude and/or effectiveness of mitigation over time exists, monitoring programmes should be proposed to enable subsequent adjustment of mitigation measures as necessary.

4   Communication of results

4.1   Presentation

The report should be laid out clearly with the minimum amount of technical terms. An index, glossary and full references should be given and the information presented so as to be comprehensible to the non-specialist.

4.2   Balance

The environmental statement should be an independent objective assessment of environmental impacts not a best case statement for the development. Negative impacts should be given equal prominence with positive impacts, and adverse impacts should not be disguised by euphemisms or platitudes. Prominence and emphasis should be given to predict large negative or positive impacts.

4.3   Non-technical summary

There should be a non-technical summary outlining the main conclusions and how they were reached. The summary should be comprehensive, containing at least a brief description of the project and the environment, an account of the main mitigating measures to be undertaken by the developer, and a description of any remaining or residual impacts. A brief explanation of the methods by which these data were obtained and an indication of the confidence which can be placed in them should also be included.

Appendix Five

List of environmental assessment regulations

i	Town and Country Planning (Assessment of Environmental Effects) Regulations 1988 (SI No. 1199)
ii	Environmental Assessment Scotland Regulations 1988 (SI No. 1221)
iii	Environmental Assessment (Scotland) Regulations 1988 (SI No. 1218)
iv	Environmental Assessment (Afforestation) Regulations 1988 (SI No. 1207)
v	Land Drainage Improvement Works (Assessment of Environmental Effects) Regulations 1988 (SI No. 1217)
vi	Highways (Assessment of Environmental Effects) Regulations 1988 (SI No. 1241)
vii	Harbour Works (Assessment of Environmental Effects) Regulations 1988 (SI No. 1336)
viii	Town and Country Planning General Development (Amendment) Order 1988 (SI No. 1272). *Note:* revoked by 1988 No. 1813 (The Town and Country Planning General Development Order 1988). Provisions of SI 1988 No. 1272 now form Article 14(2) of the 1988 General Development Order.
ix	Town and Country Planning (General Development) (Scotland) Amendment Order 1988 (SI No. 977)
x	Town and Country Planning (General Development) (Scotland) Amendment No. 2 Order 1988 (SI No. 1249)
xi	Electricity and Pipe-line Works (Assessment of Environmental Effects) Regulations 1989 (SI No. 167). *Note:* revoked by SI 1990 No. 442 (see item xiv below).
xii	Harbour Works (Assessment of Environmental Effects) (No. 2) Regulations 1989 (SI No. 424)
xiii	Town and Country Planning (Assessment of Environmental Effects) (Amendment) Regulations 1990 (SI No. 367)
xiv	Electricity and Pipe-line Works (Assessment of Environmental Effects) Regulations 1990 (SI No. 442)
xv	Roads (Assessment of Environmental Effects) Regulations (Northern Ireland) 1988 (SR No. 344)
xvi	Planning (Assessment of Environmental Effects) Regulations (Northern Ireland) 1989 (SR No. 20)
xvii	Environmental Assessment (Afforestation) Regulations (Northern Ireland) 1989 (SR No. 226)
xviii	Harbour Works (Assessment of Environmental Effects) Regulations (Northern Ireland) 1990 (SR No. 181)

Appendix Six

Guidelines on photomontage and
CAD

A photomontage is the superimposition of an image onto a photo-graph for the purpose of creating a realistic representation of pro-posed or potential changes to a view. This can be done manually by hand rendering, or by using computer imagery.

Photomontages are prepared as follows:

1   Field photograph of development site taken from fixed view-point

35mm photography is generally suitable for most developments although prints larger than A4 size lose detail. For high quality images and larger prints, 5 x 4 (inches) format is preferred. The for-mat of the photograph and the focal length should be noted and should normally be consistent between different views.

The viewpoint must be fixed, either from a known Ordnance Datum location and level, or by surveying the camera position to provide a precise coordinate. The angle of direction through the centre of the lens should also be recorded, as should be the height of the camera above ground level.

The location and height of at least three reference points in the pho-tographic view should be recorded or surveyed. These might include a visible building or transmission tower, or a known triangulation point, height of landform or landmark.

2   Preparation of geometric perspective based on available information for proposals

The degree of detail in the montage will very much depend upon that available in the proposals. For all development, basic dimen-sions of buildings, structures or landform will be required, as will information of colour finishes and landscape design elements.

3   Superimposition of perspective image onto a base photo-graph and rendering (black and white or colour) of that image to produce the photomontage

High quality hand rendered montages, where the perspective has been accurately and geometrically set up, are of considerable value in presentation. The use of the computer technique of photomon-tage allows ready incorporation of future image modifications and can rapidly be revised as the scheme is 'firmed up'. They can also later be used to test the visual impact of alternative layouts and development form and grouping.

4   Photographing of photomontage to provide reproducible copy (prints or slides)

The photograph of the photomontage should match as closely as possible the 'before' photograph in colour, clarity, *scale* and appearance. There is a tendency for the photomontage to appear darker than the 'before' print.

Data Input

OS Digital Data may be obtained at different scales for inputting topographical information into a computer. 1:50,000 series is suitable for most visual impact assessment projects with 10m contour intervals. 5m contour intervals are better for flatter sites and these sheets can be ordered from the OS. It may be necessary for accuracy, to digitise additional data, for example additional spot heights (ie. peaks, and valleys) and additional contours in the vicinity of the site from more detailed maps. An OS Digital Licence is required to hold data and copyright permission is required from OS to digitise additional data. It is important to note that any changes to purchased OS data should always be explained and clarified, as they may either increase accuracy, or introduce human error.

Stereoscopic height data from aerial surveys is very useful, particularly where access may be restricted for conventional surveying or to obtain heights of reference points, such as transmission towers or chimneys. Several specialist companies provide such a service at different scales and can provide a disc with the required data. It is important to base on the most up to date photographs and verify information on site. It is also essential to state the degree of accuracy obtained, for example ±300mm.

Visibility Mapping

This can be carried out quickly and quite accurately using Computer Aided Design (CAD) software. It must be remembered that the output is as accurate as the data used (see Data Input). Zone of Visual Influence (ZVI) or Visual Envelopes of any development can be refined by inputting elements such as buildings in the terrain model of an area. It is most useful to plot out ZVIs as overlays over OS maps at an appropriate scale. This enables the potential visibility of different options to be compared, allows the study area to be refined and potential viewpoints to be identified. ZVIs can also be generated to show where part of a development may be visible from. This refinement can indicate, for example, where the whole structure such as a wind turbine, or just the blades, may be visible or the percentage visibility of a structure, such as a stack. If an existing structure such as a transmission line is being replaced or upgraded, a ZVI should first be produced for the existing structure as part of the baseline information, to allow a comparison to be made.

The final presentation output is most useful if plotted on an original OS base map. A composite ZVI may be used, for example, in a wind farm combining individuals ZVI for each wind turbine. In such a case, it must be made clear that the visual envelope indicates areas where all or part of the development may be seen. Some software programmes can provide visibility maps which indicate the numbers of structures such as transmission towers or wind turbines, ie. 1-10, or 10-20, which may be visible.

Once a terrain model is completed, the computer can be used to generate sections very quickly. These can be used to verify the visibility of part of a development or the intervisibility of different schemes, such as wind farms. Sections can also be included in this way as part of the illustration material.

Visualisation

Computer generated perspectives are very helpful in assessing development options at an early stage in the assessment/design process. Acetate overlays are very useful if plotted out at the same scale as the photograph. The co-ordinates and height of several visible reference points are essential for overlays and photomontage preparation. Presentation plots should be generated using the finest pen size, preferably in colour, to identify clearly different elements such as, lines representing the terrain model, reference points as well as the development proposals. The eye level of the viewer may be adjusted if required for example, to test potential views from upper floors of buildings.

Photomontages can be carried our either manually using a CAD-generated perspective or entirely by computer which is generally more expensive at the very higher resolution required for quality work. Colour is preferable to black and white, as this is more realistic.

Manual photomontages can be very effective and accurate. Considerable artistic skill is required to achieve a realistic image. This may require the blending of overlapping images, the removal of elements and addition of planting. Matt photographs are most suitable for this purpose.

Computer generated montages can be produced based on scanning photographs. Photographs should be scanned at a minimum resolution of 400 dots/inch to get an adequate clarity of image. Best results with current technology have been achieved using matt prints. Considerable technical skill is also required to obtain realistic colour rendering and shading and remove and add elements. It is easier for example, to examine different options or revise a scheme using computer generated montages and assess different colours and materials.

It also easier with scanning images to achieve a similar "before" and "after" match. Hard copy outputs are still quite expensive to reproduce and the image is often of a lower quality than the screen image. Whatever method of photomontage used, a realistic print size should be used, A3 or A4 are commonly used and guidance should be given on the distance to be held from the eye. For linear projects, such as transmission lines, fold out montages may be necessary. Split montages although cheaper and easier to produce, should be avoided.

It is also important to agree and state the apparent timing of the image ie, on completion of the project with 5 year planting. This is particularly important if a project goes to public inquiry. Where possible, add as much detail as possible to achieve a realistic image. This may include the plume from a cooling tower, a transmission line from a wind farm or access road and minor buildings.

Although quite costly to produce, it is well worth considering the use of videomontages to cover moving elements such as wind turbines. These should be professionally produced with a voice over describing the view. This could be considered for closed viewpoints and 5-10 minutes should be adequate, including panning shots to show the landscape context. As with photomontages, adequate detail is required to show ancillary development, whether roads, power lines or sub-stations, in the case of wind farms.

Index